Other Avon Books by
Isaac Asimov

THE ROAD TO INFINITY

ISAAC ASIMOV

A DISCUS BOOK/PUBLISHED BY AVON BOOKS

The following essays in this volume are reprinted from *The Magazine of Fantasy and Science Fiction*, having appeared in the indicated issues:

The Subtlest Difference, October 1977
The Sons of Mars Revisited, November 1977
Dark and Bright, December 1977
The Real Finds Waiting, January 1978
The Lost Art, February 1978
Anyone for Tens? March 1978
The Floating Crystal Palace, April 1978
By Land and By Sea, May 1978
We Were the First That Ever Burst—, June 1978
Second to the Skua, July 1978
Rings and Things, August 1978
Countdown, September 1978
Toward Zero, October 1978
Fifty Million Big Brothers, November 1978
Where Is Everybody? December 1978
Proxima, January 1979
The Road to Infinity, February 1979

AVON BOOKS
A division of
The Hearst Corporation
959 Eighth Avenue
New York, New York 10019

First Avon Printing, February, 1981

Dedicated to

The Men and Women of Mensa,
who make the best audience in the world.

Contents

Introduction   ix

A   Numbers

1   The Lost Art   3
2   Anyone for Tens?   16

B   Elements

3   Countdown   33
4   Toward Zero   45

C   Earth

5   The Floating Crystal Palace   59
6   By Land and By Sea   72
7   We Were the First That Ever Burst—   85
8   Second to the Skua   98

D   Planets

9   The Sons of Mars Revisited   113
10   Dark and Bright   125
11   The Real Finds Waiting   137
12   Rings and Things   150

E   Stars

13   Proxima   165

CONTENTS

F   Universe

14   Fifty Million Big Brothers   181
15   Where Is Everybody?   193
16   The Road to Infinity   206

G   Life and Death

17   The Subtlest Difference   221

Introduction

One of the essays in this book, "Fifty Million Big Brothers," appeared in the November 1978 issue of *The Magazine of Fantasy and Science Fiction* (usually known as *F & SF*). It was one of a monthly series of such essays, the first of which appeared in the November 1958 issue of *F & SF*.

With "Fifty Million Big Brothers," I celebrated the twentieth anniversary of the series, started my twenty-first year—and kept right on going. At the time I am writing this I have completed my 244th *F & SF* essay, and of the 244, 237 (plus a 238th from another source) have appeared in one or another of a whole series of collections published by the long-suffering firm of Doubleday and Company, Inc.

If you don't mind, then, I'm going to have to pause in order to catch my breath. Every once in a while, a Gentle Reader writes to ask whether I have ever dealt with a particular subject in one of my essays. Every once in another while, another Gentle Reader describes an essay of mine and asks where he can find it.

And, most important of all, this Harried Writer is beginning to have trouble keeping them all straight in his mind. I live in continual fear that I will do an article with great satisfaction and then be informed by hordes of Ungentle Ex-readers that I had repeated an ex-article, paragraph by paragraph.

Perhaps, then, you won't mind if, instead of my usual fascinating introduction, I subject you to a list of the 238 collected essays in alphabetical order, together with a few words on the subject of each and a code number that will give you the particular collection in which it appears.

In that way, I celebrate the twentieth anniversary of the series, I oblige various Gentle Readers, and I have a tool for keeping myself on the track—if it works.

LIST

Title	Collection	Subject
Academe and I	10	personal reminiscences
A Galaxy at a Time	4, 14	galactic explosions
All Gall	12	cholesterol
A Matter of Scale	6	scale of the universe
Ancient and the Ultimate	10	future of books
Anyone for Tens?	18	logarithms
A Particular Matter	11	natural radioactivity
A Piece of Pi	3, 17	pi
A Piece of the Action	4, 16	Planck's constant
As Easy as Two Plus Three	11	hydrogen fusion
Asimov's Corollary	13	popular fallacies
Asymmetry of Life	9	stereoisomerism
At Closest Range	11	tissue radioactivity
Backward, Turn Backward	7	planetary rotations
Balancing the Books	6	conservation laws
Battle of the Eggheads	1	intellectual bigotry
BB or Not BB, That Is the Question	6	origin of the universe
Begin at the Beginning	4, 17	counting the years
Behind the Teacher's Back	5, 16	uncertainty principle
Best Foot Backward	12	importance of technology
Beyond Pluto	1, 14	tenth planet
Bill and I	9	Shakespeare's astronomy
Black of Night	4, 14	expanding universe
Bridge of the Gods	12	light spectra

INTRODUCTION

Title	Collection	Subject
Bridging the Gaps	8	periodic table of the elements
But How?	9	population problem
By Jove	2, 14	planet Jupiter
By Land and by Sea	18	Antarctic exploration
By the Numbers	10	computers
Catching Up with Newton	1, 16	escape velocities
Catskills in the Sky	1	views from the planets
Certainty of Uncertainty	5, 16	uncertainty principle
C for Celeritas	4, 16	speed of light
Change of Air	12	fluorocarbons; ozone layer
Cinderella Compound	10	nucleic acids
Clock in the Sky	10	Jupiter's satellites
Cold Water	9	freezing of water
Comet That Wasn't	13	discovery of Uranus
Constant as the Northern Star	11	precession of the equinoxes
Countdown	18	liquefaction of gases
Counting Chromosomes	7	x- and y-chromosomes
Crowded!	6	city populations
Cruise and I	10	personal reminiscences
Dance of the Luminaries	11	solar eclipses
Dance of the Satellites	7	Jupiter's satellites
Dance of the Sun	7	Sun's motion across the sky
Dark and Bright	18	satellites of Mars
Dark Companion	13	white dwarfs
Days of Our Years	4, 17	various calendars
Death in the Laboratory	5, 15	fluorine
Discovery by Blink	13	discovery of Pluto

INTRODUCTION

Title	Collection	Subject
Distance of Far	8	recession of the galaxies
Doctor, Doctor, Cut My Throat	10	thyroid gland
Double-ended Candle	11	American energy policy
Down from the Amoeba	10	bacteria and viruses
Dying Lizards	7	dinosaur extinctions
Eclipse and I	11	solar eclipses
Egg and Wee	2, 15	cell size
Element of Perfection	2, 15	helium
Euclid's Fifth	9	geometric axioms
Eureka Phenomenon	9	scientific inspiration
Evens Have It	2, 15	nuclear structure
Exclamation Point!	5, 17	factorial numbers
Fateful Lightning	8	Franklin's lightning rod
Fifty Million Big Brothers	18	extraterrestrial intelligence
Figure of the Farthest	11	size of the universe
Figure of the Fastest	11	speed of light
Final Collapse	13	black holes
First and Rearmost	4, 16	gravitation
First Metal	7	metals known to the ancients
Flickering Yardstick	1, 14	Cepheid variables
Floating Crystal Palace	18	icebergs
Forget It!	4, 17	old measurement systems
Future? Tense!	5	futurology
Ghost Lines in the Sky	4	latitude and longitude
Great Borning	6	Cambrian fossils

INTRODUCTION

Title	Collection	Subject
Harmony in Heaven	5, 14	Kepler's third law
Haste-makers	4, 15	enzymes
Heavenly Zoo	4	zodiac
Heaven on Earth	1, 14	mapping the stars
Height of Up	2, 16	temperature extremes
Here It Comes; There It Goes	1	continuous creation
He's Not My Type	3	blood types
Holes in the Head	9	duckbill platypus
Hot Stuff	2	neutrinos and super-novas
Hot Water	9	boiling points
Imaginary That Isn't	3, 17	imaginary numbers
I'm Looking over a Four-leaf Clover	6	origin of the universe
Impossible, That's All	6	speed of light
Incredible Shrinking People	7	miniaturization
Inevitability of Life	11	origin of life
Isaac Winners	3, 15	great scientists in history
Isles of Earth	5, 17	islands
It's a Wonderful Town	13	New York City
Judo Argument	12	existence of God
Just Mooning Around	4, 14	satellites
Just Right	7	square-cube law
Kaleidoscope in the Sky	6	satellites of Mars
Knock Plastic	6	popular fallacies
Land of Mu	5, 16	mesons
Left Hand of the Electron	9	conservation of parity

Title	Collection	Subject
Life's Bottleneck	1, 15	elements in living tissue
Light Fantastic	3, 16	lasers
Light That Failed	3, 16	Michelson-Morley experiment
Little Found Satellite	7	Saturn and its rings
Little Lost Satellite	7	large asteroids
Look Long upon a Monkey	11	apes and men
Lopsided Sun	8	tidal influences on sun
Lost Art	18	logarithms
Lost Generation	3	Mendelian laws
Lost in Non-translation	10	Book of Ruth
Lunar Honor Roll	8	mapping the Moon
Luxon Wall	8	tachyons
Magic Isle	13	undiscovered elements
Making It	13	American technology
Man Who Massed the Earth	8	mass of the Earth
Mispronounced Metal	11	aluminum
Modern Demonology	2, 16	Maxwell's demon
Moon over Babylon	10	week
Moving Ahead	13	American technology
Multiplying Elements	8	rare-earth metals
Music to my Ears	6	musical scale
My Built-in Doubter	1	popular fallacies
My Planet, 'Tis of Thee	8	world government
Nightfall Effect	12	space settlements
Nobelmen of Science	5, 15	great scientists
Nobel Prize That Wasn't	8	atomic numbers
No More Ice Ages?	1, 15	greenhouse effect

INTRODUCTION

Title	Collection	Subject
Non-time Travel	6	international date line
Not as We Know It	2, 15	chemical bases of life
Nothing	7	interstellar dust
Nothing Counts	4, 17	discovery of zero
Now Hear This	2, 16	echo location
Oblique the Central Globe	13	axial tipping
Odds and Evens	9	conservation of parity
Of Capture and Escape	1, 16	escape velocity
Of Ice and Man	13	orbital eccentricity
Oh, East Is West and West Is East	5	latitude and longitude
O Keen-eyed Peerer into the Future	11	futurology
Old Man River	6	rivers
Olympian Snows	12	surface of Mars
One and Only	10	carbon
One, Ten, Buckle My Shoe	3, 17	binary numbers
On Throwing a Ball	8	gravitation
Opposite Poles	13	ice ages
Order! Order!	2, 16	entropy
Planetary Eccentric	7	Pluto
Planet of the Double Sun	1, 14	Alpha Centauri
Plane Truth	9	non-Euclidean geometry
Planet That Wasn't	12	Vulcan
Playing the Game	8	Doppler-Fizeau effect
Pompey and Circumstance	9	role of coincidence
Portrait of the Writer as a Boy	6	personal reminiscence
Power of Progression	8	overpopulation

Title	Collection	Subject
Predicted Metal	7	periodic table
Pre-fixing It Up	3, 17	metric system
Prime Quality	9	prime numbers
Proton-Reckoner	5, 17	large numbers
Proxima	18	Sun's companion
Quasar, Quasar, Burning Bright	13	luminosity of objects in sky
Real Finds Waiting	18	satellites of Mars
Recipe for a Planet	2, 15	elementary makeup of Earth
Right Beneath Your Feet	6	antipodes
Rigid Vacuum	3, 16	luminiferous ether
Rings and Things	18	Uranus' rings; Chiron
Road to Infinity	18	size of black holes
Rocketing Dutchmen	12	flying saucers
Rocks of Damocles	5, 14	Earth-grazers
Roll Call	4	planetary nomen-clature
Round and Round and—	4, 14	planetary rotations
Sea-green Planet	13	discovery of Neptune
Second to the Skua	18	Antarctica
Seeing Double	9	light polarization
Seventh Metal	7	element mercury
Seventh Planet	7	Venus and Mercury
Shape of Things	3	shape of the Earth
Sight of Home	1, 14	luminosity of the stars
Signs of the Times	11	zodiac and preces-sion
Silent Victory	12	oxygen atmosphere
Sin of the Scientist	8	poison gas
Skewered!	11	large numbers

Title	Collection	Subject
Slow Burn	3, 15	oxygen and oxidation
Slowly Moving Finger	4	aging
Smell of Electricity	12	ozone
Sons of Mars Revisited	18	satellites of Mars
Squ-u-u-ush!	5	collapsed stars
Star in the East	12	Star of Bethlehem
Stars in Their Courses	8	astrology
Stepping Stones to the Stars	1, 14	comets
Stop!	9	birthrate
Subtlest Difference	18	life and death
Superficially Speaking	2, 14	planetary areas
Surprise! Surprise!	13	undiscovered elements
Symbol-minded Chemist	6	atomic symbols
Terrible Lizards	7	dinosaurs
T-Formation	3, 17	large numbers
Thalassogens	9	common liquids
That's About the Size of It	2, 17	animal extremes of size
That's Life	2, 15	life and non-life
Thin Air	1, 16	upper atmosphere
Thinking about Thinking	12	IQ tests and intelligence
Third Liquid	12	melting points of elements
Those Crazy Ideas	1	creativity
Three-D Molecule	9	molecular structure
Through the Microglass	10	micro-organisms
Time and Tide	5, 14	tidal effects
Times of Our Lives	6	time-zones
Titanic Surprise	12	world sizes in solar system
Tools of the Trade	3, 17	pi
To Tell a Chemist	5, 15	Avogadro's number

INTRODUCTION

Title	Collection	Subject
To the Top	13	American technology
Toward Zero	18	absolute zero
Tragedy of the Moon	10	ancient astronomic theory
Triumph of the Moon	10	influence of Moon on life
Trojan Hearse	2, 14	Trojan asteroids
Twelve Point Three Six Nine	6	popular fallacies
Twinkle, Twinkle, Little Star	3, 14	white dwarfs
Twinkle, Twinkle, Microwaves	13	discovery of pulsars
Two at a Time	8	perturbations
Ultimate Split of the Second	2, 16	ultrashort time periods
Uncertain, Coy, and Hard to Please	7	women's rights
Uneternal Atoms	11	natural radioactivity
Unlikely Twins	10	graphite and diamond
Up and Down the Earth	5, 17	mountains and deeps
Updating the Asteroids	11	Earth-grazers
Varieties of the Infinite	3, 17	transfinite numbers
View from Amalthea	7	Jupiter's satellites
Water, Water, Everywhere	5, 17	oceans
Week Excuse	10	calendar reform
Weighting Game	2, 15	atomic weights
Welcome, Stranger	4, 15	noble gas compounds
We Were the First That Ever Burst—	18	oceanic exploration

INTRODUCTION

Title	Collection	Subject
Where Is Everybody?	18	extraterrestrial intelligence
Wicked Witch Is Dead	12	women and aging
World: Ceres	10	large asteroids
Worlds in Confusion	8	Velikovsky's notions
Wrong Turning	12	retrograde satellites
You, Too, Can Speak Gaelic	3, 15	chemical nomenclature

Key to Location of Essays in Collections

1 Fact and Fancy
2 View from a Height
3 Adding a Dimension
4 Of Time and Space and Other Things
5 From Earth to Heaven
6 Science, Numbers, and I
7 The Solar System and Back
8 The Stars in Their Courses
9 The Left Hand of the Electron
10 The Tragedy of the Moon
11 Of Matters Great and Small
12 The Planet That Wasn't
13 Quasar, Quasar, Burning Bright
14 Asimov on Astronomy
15 Asimov on Chemistry
16 Asimov on Physics
17 Asimov on Numbers
18 The Road to Infinity

A NUMBERS

1. The Lost Art

It is one thing to be able to make predictions. It is another to listen to the predictions you have made and to act upon them.

For instance (just so that my Gentle Readers will know what I'm talking about), back in 1950 I wrote a short story that was to serve as an introductory section to the first book of *The Foundation Trilogy*, which was then soon to be published.

In it, the great psychohistorian, Hari Seldon, is about to demonstrate an important point. Here is the passage:

"Seldon removed his calculator pad from the pouch at his belt. Men said he kept one beneath his pillow for use in moments of wakefulness. Its gray glossy finish was slightly worn by use. Seldon's nimble fingers, spotted now with age, played along the hard plastic that rimmed it. Red symbols glowed out from the gray."

There! That's not bad as a prediction of the pocket computer that is now so ubiquitous. I even got the color of the symbols right.

Seven years later, in 1957, I wrote a short story called "The Feeling of Power," in which I described a society in which computers had become so ubiquitous that pen-and-paper calculation became a lost art. In the story, my hero claims to have rediscovered the art. He announces that nine times seven is sixty-three. The congressman who hears this doesn't believe it—

"The congressman took out his pocket computer, nudged

the milled edges twice, looked at its face as it lay there in the palm of his hand—"

Another prediction of the pocket computer (which, by the way, agreed that nine times seven is sixty-three). I even got the name right.

Another seven years passed and I decided to write a book on the slide rule. I finished it in February 1965. I called it "An Easy Introduction to the Slide Rule" and it was published by Houghton Mifflin toward the end of the year.

But by that time the pocket computers I had myself predicted a decade and a half before were coming into use and growing cheaper every year. In no time at all, the use of the slide rule had become a lost art and my book was lost effort. They don't even *manufacture* slide rules any more. —See? I make the prediction and then I don't listen to it.

The slide rule is only a mechanical device for handling logarithms. You can get the same results to more decimal places (albeit more tediously) with a good table of logarithms (or "log table," for short). And now the use of logarithms is a lost art, too.

You might think, then, that it is ridiculous of me, after the fact, to devote space to logarithms. But I'm going to do it anyway, for three reasons:

1. I want to.
2. You might find it interesting. After all, things don't necessarily have to be useful to be interesting.
3. It might come in handy. Suppose you have been stranded on a desert island and the batteries in your pocket computer go dead. If you must then make some complicated computation you needn't be helpless, provided you understand logarithms. You just chop down a tree and construct a log table.*

Logarithms were invented in 1614 by a Scottish mathematician named John Napier, who worked twenty years to perfect the notion. His original motive was to find a method

* If any of you cynics out there think I'm writing this article just to get this sentence into print—you may be right.

4

for reducing the load of work required in tedious computations involving certain trigonometric functions called "sines," these computations being important in astronomical work.

A sine is a ratio. In a right triangle, the sine of one of the acute angles is the ratio of the length of the side opposite the angle to the length of the hypotenuse. Because of this, Napier named the numbers, invented to help in computing these ratios, "logarithms" from Greek words meaning "ratio number."

It was only afterward that he realized that logarithms were useful in tedious computations involving *any* numbers. Their usefulness was not confined to these particular ratios, and the word "logarithm" is therefore a misnomer—but we are stuck with it.†

Now that that is clear, let's forget about Napier and make a fresh start.

I think we'll all admit that multiplication and division are much harder than addition and subtraction. Even those prodigies among you who can multiply and divide very quickly and who get the right answer eight or nine times out of ten, and who therefore think it's easy, will have to admit that addition and subtraction are easier still.

If, then, anyone can discover a way of converting a multiplication into an addition and a division into a subtraction, he deserves the plaudits of the multitude. Actually, it can be done.

Take the problem 8×7. This is shorthand for the question "What is the sum of eight 7s?" Since the multiplication, in ordinary arithmetic, is commutative, $8 \times 7 = 7 \times 8$, and the latter is shorthand for "What is the sum of seven 8s?"

Either sum is child's play, since $7+7+7+7+7+7+7+7=56$, and $8+8+8+8+8+8+8=56$.

---

† When I was young, I came across the following riddle: Why is a woodcutters' ball like a mathematics textbook? —The answer is: Because both are full of logger rhythms.

In the same way, the problem $28 \div 7$ is a shorthand way of saying "How many times can you subtract 7 from 28 before you reach zero?" Since $28-7-7-7-7=0$, the clear answer is $28 \div 7=4$.

Of course, if you had as your problem $30 \div 7$, you are in a quandary. Subtract four 7s and you haven't reached 0; subtract five 7s and you've passed 0. The answer is to get as close to 0 as you can, without passing it and then subtract whatever number is required to get you to zero. Thus, $30-7-7-7-7-2=0$. You can then say that $30 \div 7=4$, with 2 left over; or, with a little more sophistication of notation, $30 \div 7=4 \ 2/7$.

Fine! We have now converted multiplication into addition and division into subtraction and we are happy as can be—as long as we stay with small numbers.

What happens, though, if we want to solve 3498×729? Adding three thousand four hundred ninety-eight 729s or seven hundred twenty-nine 3498s is going to be a terrible chore, and I, for one, refuse to tackle it. Trying to do $75,643 \div 803$ by the subtraction method is even worse.

So, as a matter of fact, schoolchildren are not taught to multiply and divide by adding and subtracting. They are, instead, taught the multiplication and division of small numbers by drill and brute memory. The average youngster knows that $7 \times 8=56$ and that $28 \div 7=4$ simply because he has memorized all the multiplications from 0×0 to perhaps as high as 12×12, and all the divisions from $144 \div 12$ to $0 \div 1$. (Sometimes he even remembers some of the multiplications and divisions into adult life.)

Then, for higher numbers still, the youngster learns various tricks that involve the simple memorized products and quotients together with a mixed bag of indenting and carrying numbers that not one child in a hundred knows the reasons for. You just learn it as a fixed set of rules like those involved in driving a car, with this difference—no one who learns how to drive a car ever forgets how.

Next, let's start all over again with something that may seem to have nothing to do with the problem. We'll write a series of numbers beginning with 2 and double it each time, thus: 2, 4, 8, 16, 32, 64, 128, 256, 512, 1024, 2048, 4096, 8192. . . . You can continue it as far as you like.

This series has an interesting property. —Suppose you multiply any two numbers in the series or for that matter any number in that series by itself, thus: $8 \times 32 = 256$, or $64 \times 64 = 4096$. You'll notice that the products, 256 and 4096, are also members of the series. This is always true. You can try as many examples as you wish. All multiplications of members in the series (involving two, three, or any finite number of members of this series) will end by giving you a product which is a member of the series—if you extend the series far enough.

The series we have constructed is, in other words, "closed to multiplication."

Why is this so? If we write the series in another fashion the answer becomes obvious. The first number of the series is 2; the second number is twice the first, or 2×2 (which is 4). The third number is twice the second, or $2 \times 2 \times 2$ (which is 8). The fourth number is twice the third, or $2 \times 2 \times 2 \times 2$ (which is 16). —And so on forever.

If we write the series as multiples of 2, it will become: $2; 2 \times 2; 2 \times 2 \times 2; 2 \times 2 \times 2 \times 2; 2 \times 2 \times 2 \times 2 \times 2; \ldots$

This is an infinite series made up of every number in existence from 2 on up that can be constructed by multiplying 2s. Any number that can be so constructed is on the list somewhere.

If, then, we want to solve 8×32 and write it as multiples of 2, we are trying to work out $(2 \times 2 \times 2) \times (2 \times 2 \times 2 \times 2 \times 2)$. The answer is $2 \times 2 \times 2 \times 2 \times 2 \times 2 \times 2 \times 2$, or 256. The product, being a product of 2s, is inevitably part of the series. Multiply any product of a given number of 2s by any other product of a given number of 2s and the over-all product is *still* a product of a given number of 2s.

Notice, also, that in the problem we have just considered $8 \times 32 = 256$, the product of three 2s multiplied by the product of five 2s gave an over-all product that is the product of eight 2s. You may be multiplying the products, but you're *adding* $(3+5=8)$ the number of 2s being multiplied.

This becomes clearer and more useful if we stop trying to write all the 2s, since that is not only messy but it makes it easy to miss count and make mistakes. Instead, let's write the multiplication of two 2s as 2^2, of three 2s as 2^3, of four 2s as 2^4, and so on. The 2 written in the ordinary fashion is the "base" and the smaller number to the upper right is the "exponent." A number like 2^4 is an "exponential number."

The number 2^2 is usually read "2 square" and 2^3 is read "2 cube." From there on in, we say "2 to the fourth power," "2 to the fifth power," and so on. My own feeling is that it would be much better to drop the "square" and "cube," which go back to the geometric predilections of the ancient Greeks and make the notation uniform by speaking of "2 to the second power" and "2 to the third power."

The number 2 itself is just one 2, and you can't really speak of the "multiplication of one 2," but, for uniformity's sake, you might as well write 2 as 2^1, if you wish to put it into exponential form, and call it "2 to the first power."

Our original series, 2, 4, 8, 16, 32 . . . , can therefore be written, instead, as 2^1, 2^2, 2^3, 2^4, 2^5. . . . You have to admit that using exponential numbers is a neater way of presenting the series.

Instead of writing $8 \times 32 = 256$, then, we could write the problem exponentially, as $2^3 \times 2^5 = 2^8$. Notice again that although we are multiplying 2^3 and 2^5, we are *adding* the exponents 3 and 5.

Suppose we prepare a table such as Table 1.

Using Table 1, it would be very easy to solve a problem such as 256×8192. Using the table (which we can imagine being extended to huge numbers in the trillions of trillions), we would find that for 256 the exponent to the base 2 is 8 and for 8192 it is 13.

Instead of multiplying, we would simply add the exponents: $8+13=21$. We would then use the table to find the exponent 21 and look to the left to see the number it repre-

TABLE 1

Number	Exponent to the base 2
2	1
4	2
8	3
16	4
32	5
64	6
128	7
256	8
512	9
1024	10
2048	11
4096	12
8192	13
and so on	and so on

sents. That number is 2,097,152. We therefore would know that $256 \times 8192 = 2,097,152$.

As it happens, in tables such as Table 1, the right-hand column is not usually labeled "*Exponent (to the base 2)*," but "*Logarithm (to the base 2)*." This is pure semantics. A logarithm *is* an exponent and I'm sorry Napier invented the word. Now both words exist and add to the confusion in the world. The left-hand column in Table 1 can be headed "*Antilogarithm*" rather than "*Number.*" (To save syllables, I will speak of "logs" and "antilogs.")

Thus we can say "$\log_2 4 = 2$" (or "the log of 4 to the base 2 is 2"). Similarly, "$\log_2 8 = 3$," "$\log_2 16 = 4$," and so on indefinitely. What we have done in multiplying 256 and 8192 is to make use of a log table.

Can we also divide by using exponents (or logs)? Suppose you want to divide 64 by 16. It works out to $64 \div 16 = 4$, or, in exponential notation, $2^6 \div 2^4 = 2^2$.

We can see at a glance that division seems to be involved with the subtraction of exponents, which is no surprise since division is the opposite of multiplication. If multiplication involves the addition of exponents, then division *should* involve the subtraction of exponents.

Suppose, then, we try to solve the following problem: 8,388,608÷524,288. We can look up the logarithms in Table 1 (assuming it to be extended to large enough figures) and find that $\log_2 8,388,608 = 23$ and $\log_2 524,288 = 19$. We subtract the logs—$23 - 19 = 4$—and that means the answer is that number whose log to the base 2 is 4. That number is 16, since $\log_2 16 = 4$, as Table 1 tells us (if we haven't already memorized the fact). Now we know that $8,388,608 ÷ 524,288 = 16$, and all we've done is use a log table and subtract. No division at all.

But division isn't quite that easy. The series of numbers that includes all those made up of 2s multiplied together is closed to multiplication, in that *any* number in the series multiplied by *any other* in the series yields a product that is also a number in the series. This is not true the other way around; the series as I have given it is *not* closed to division.

For instance $4÷4=1$, and though 4 is a member of the series, 1 is not. Again $2÷16=1/8$, and though both 2 and 16 are members of the series, 1/8 is not.

Suppose, then, we extend the series. We had started arbitrarily with 2 and had then kept doubling and redoubling to build up larger and larger numbers without limit. But if we are going to be limitless at one end, why not limitless at the other?

Suppose we start at 2 again and this time halve and halve again to produce smaller and smaller numbers? Thus 2, 1, 1/2, 1/4, 1/8, 1/16, 1/32, 1/64 . . . , and so on endlessly.

The new series, working both ways from 2, would be endless in both directions and the central part of it would look like this if we move from the very small on the left to the very large on the right: . . . 1/64, 1/32, 1/16, 1/8, 1/4, 1/2, 1, 2, 4, 8, 16, 32, 64. . . .

Such as extended series is closed to *both* multiplication and division. Thus $1/8 × 1/4 = 1/32$ and $8÷32 = 1/4$ and $1/4÷1/16 = 4$, and $2×4÷16×1/2÷1/8 = 2$, and so on.

You can take any finite number of members from the extended series, with or without repetition, and make use of any finite number of multiplications and divisions involving them, and the result will always remain a member of the series.

The next step is to work out exponential notation for the extended series.

If we grant that $4=2^2$ and $2=2^1$, what is 1 equal to in exponential notation? Well, $4 \div 4$ or $2 \div 2$ (or, for that matter, any number divided by itself) is equal to 1. That means that $2^2 \div 2^2 = 1$ or $2^1 \div 2^1 = 1$ (as does any other exponential number divided by itself). If division involves the subtraction of exponents, then $2^2 \div 2^2 = 2^0$, since $2-2=0$, and this is true for any exponential number divided by itself.

The only way a particular division can have two different answers is when the two different answers are really the same, so that we must define 2^0 as being equal to 1.‡ This doesn't *seem* to make sense since 2^0 would appear to mean zero 2s multiplied together and that sounds, intuitively, as though it should equal 0. However, to keep exponential notation self-consistent (and the great basic rule in mathematics is that any mathematical system must be self-consistent), 2^0 must be set equal to 1.

Then, what about 1/2? Since $4 \div 8 = 1/2$, then $2^2 \div 2^3 = 1/2$. By subtracting exponents, we find that $2^2 \div 2^3 = 2^{-1}$. Therefore $1/2 = 2^{-1}$. Working similarly, we find that $1/4 = 2^{-2}$, $1/8 = 2^{-3}$, $1/16 = 2^{-4}$, and so on.

There seems to be a symmetry here, and that is always a good thing to find in mathematics. Thus $1/2 = 1/2^1 = 2^{-1}$, $1/4 = 1/2^2 = 2^{-2}$, $1/8 = 1/2^3 = 2^{-3}$. Such symmetries help assure one that the system makes sense.

Now, then, we can make an extended log table as in Table 2.

Using Table 2, we can multiply and divide freely by

‡ Please don't write to tell me that any number divided by 0 has an infinite number of different answers. It is for this very reason that division by zero is excluded in mathematics.

TABLE 2

Antilog	Log to the base 2
·	·
·	·
·	·
·	·
1/256	−8
1/128	−7
1/64	−6
1/32	−5
1/16	−4
1/8	−3
1/4	−2
1/2	−1
1	0
2	1
4	2
8	3
16	4
32	5
64	6
128	7
256	8
·	·
·	·
·	·
·	·
·	·

adding and subtracting logs as long as we stick to the numbers in the series.

But (and it's an awfully big "but") it *only* works for numbers in the series, and those numbers in the series are only a vanishingly small fraction of all the numbers there are.

For instance, from 1 to 100 there are only seven whole numbers—1, 2, 4, 8, 16, 32, and 64—that are members of the series, while there are ninety-three numbers that are not. If we choose any two numbers from 1 to 100, we have a possible total of ten thousand combinations that can be multiplied or divided, and only forty-nine of those combinations will be soluble by the logs to the base 2. That's only about one out of two hundred, and the situation grows worse as the numbers grow larger. The same is also true of unit fractions (those with 1 in the numerator.)

Now what do we do?

Suppose we try other numbers as the base for a series. We might start with 3 and multiply it by 3 over and over to get larger and larger numbers, then divide it by 3 over and over to get smaller and smaller numbers. We would end up with a series like this: . . . 1/729, 1/243, 1/81, 1/9, 1/3, 1, 3, 9, 81, 243, 729 . . .

This could be written: . . . 3^{-5}, 3^{-4}, 3^{-3}, 3^{-2}, 3^{-1}, 3^0, 3^1, 3^2, 3^3, 3^4, 3^5 . . .

We could use this to get a table of logarithms to the base 3. This would give us a whole series of numbers that are not located on the 2 series. (Except for 1, that is, which is common to all such series. After all, $3^0=1$ just as $2^0=1$, and by the same reasoning. Indeed, *any* number to the zeroth power is equal to 1.)

We might go higher still. There's no use using 4 as a base because that's one of the numbers in the 2 series, and if we start with 4 and multiply it over and over again by 4, we end up with a series all members of which are included in the 2 series—4, 16, 64, 256, 1024, and so on. (I'll leave it to you to see how it works in the other direction, through 1 and unit fractions.)

We can, however, use 5, 6, and 7 as bases to give us new numbers that are on none of the other series. Just considering the whole numbers, there are 5, 25, 125, 625 . . . and 6, 36, 216, 1296 . . . and 7, 49, 343, 2401. . . . In exponential notation, these are, respectively, 5^1, 5^2, 5^3, 5^4 . . . ; 6^1, 6^2, 6^3, 6^4 . . . ; and 7^1, 7^2, 7^3, 7^4. . . . (Again, I'll leave it to you to see how it works in the other direction, through 1 and unit fractions.)

There's no use trying 8 or 9 as a base since the former

is present in the 2 series and the latter in the 3 series, but you can go on to 10, 11, 12, 13, 14, and so on, for additional series.

If you make an infinite number of such series, then every whole number and every unit fraction is on one series or another (and numbers which are not unit fractions, like 2/7, or mixed numbers, like 3 2/7—every rational number, in fact—can always be dealt with as a combination of a multiplication and division of whole numbers).

Of course, using an infinite number of series to make sure you can find every number is not really practical. On the other hand, may a finite number, say several dozen, do for all numbers between 1/100 and 100 perhaps?

No, it won't!

There's another catch. We can't mix series. For instance, suppose we want to solve 8×9, to which we know the answer is 72. The number 8 is equal to 2^3 and the number 9 to 3^2—can we add exponents? Suppose we consider $2^3 \times 3^2$. We know that $3+2=5$, but is the answer, then, 2^5 or 3^5?

Neither! It turns out that $2^5=32$ and $3^5=243$ and neither is even close to 72. We can't even split the difference in bases and suspect that $2^3 \times 3^2 = 2.5^5$, since $2.5 \times 2.5 \times 2.5 \times 2.5 \times 2.5 = 97.65625$ and that isn't 72, either.

The conclusion we must come to is that you can't mix series of numbers in using logs. You can't combine logs to a particular base with logs to another base and come out with a correct answer. *You must stick with a particular base* and never veer from it, if you are to get any correct answers at all.

Well, then, which base?

I've already said that the 2-series gives you only a small fraction of the numbers you are likely to encounter, so it's no good. Yet the 3-series gives you fewer numbers still and the 5-series still fewer. The higher the base you use to build a series, the fewer the numbers you will include on that series over a given stretch.

For instance, between 1 and 1000 the 2-series gives you ten numbers: 1, 2, 4, 8, 16, 32, 64, 128, 256, and 512.

The 12-series gives you only three numbers: 1, 12, 144. The next number in the 12-series is already 1728. The 33-series gives you only two numbers, 1 and 33, for the next number in the series is already 1089.

So although the 2-series is imperfect, it is less imperfect than any of the other series. Can we use something smaller than 2 as a base and get more numbers?

We can't use either 1 or 0 as a base because 1 to any power is 1, and 0 to any power is 0. (Try it: $1 \times 1 \times 1 \ldots$ or $0 \times 0 \times 0. \ldots$) The series would therefore be, 1, 1, 1, 1 ... or 0, 0, 0, 0 ... and either would be useless. Negative numbers would give you the same series as the positive numbers except that with negative numbers you would alternate between positive numbers and negative numbers, thus: -2, 4, -8, 16, -32, 64, and so on.

What about fractions? Suppose we start with 1.1 and keep multiplying by 1.1. We would have 1.1, 1.21, 1.331, 1.4641, 1.61051 and so on. This gives us a lot of decimal numbers and fills in the space between whole numbers as no whole number series could. We would even get more if we used 1.01, still more if we used 1.001 and so on.

But whatever base we use and however small the interval between successive numbers in the series, there are always myriads of numbers omitted for everyone included and the situation always gets worse and worse as you get up the scale. No matter how small the base, you will always get to places where you are skipping dozens of whole numbers—hundreds—thousands—millions—with each step.

Unless you can somehow fill in the intervals between successive terms in a series, no series will be useful. However, the intervals *can* be filled in, you will be glad to hear, so that we can, if we wish, use any series for the purpose, basing our choice on any grounds that we wish to use—pure whim, if that is our wish.

As it turns out, there are only two series that anyone uses, and the more common of these—the one on which the slide rule is based—is the 10-series. Why the 10-series?

Because it fits our number system beautifully—as I will show you in the next chapter.

2. Anyone for Tens?

Occasionally I will write an essay that will accept, as a matter of course, the development of the Universe and life and man and brain by evolutionary processes. This is taken quietly by my usual audience.

Not so if I reach beyond to people not ordinarily exposed to such ideas. If I make similar assumptions in articles in *TV Guide,* for instance, I rouse the Bible Belt, and I am promptly bombarded with letters on the iniquity of evolutionary ideas or of *any* notions that are postbiblical in nature (except for television sets, I presume).

At first, I would conscientiously try to send reasoned replies, and when it became clear that this was equivalent to trying to bail out the ocean with a spoon, I spent some time brooding on human folly. Then I decided that such brooding also got me nowhere, so what I do now is glance over each letter for laughs before dumping it.

My favorite recent letter, coming in response to an article on the "big bang" theory of the origin of the Universe, began as follows:

"The trouble with you scientists is that you don't *observe*. If you only took the trouble to make the simplest observations you would see at once that the Bible says, 'In the beginning God created the heaven and the earth.' "

Imagine scientists overlooking that key observation! And it is actually the first verse in the Bible! You would think it was impossible to miss.

It makes me feel sad that I must turn now to a mathematical topic concerning which there is no controversy and on which the Bible Belt makes no stand based on their superior powers of observation.

In Chapter 1 I discussed exponents and explained that since $16 = 2 \times 2 \times 2 \times 2$ (four 2s multiplied together), it could be expressed as 2^4. Similarly, $1/8 = 1/2 \times 1/2 \times 1/2$ and is therefore $(1/2)^3$, or 2^{-3}. Then, too, 2^0 (or any number to the zeroth power) is equal to 1. In a number like 2^4, the 4 is an exponent, the 2 is the base.

When two exponential numbers with the same base are multiplied, the exponents are added; when divided, the exponent of the divisor is subtracted from that of the dividend. Thus, $2^5 \times 2^3 = 2^8$ and $2^5 \div 2^3 = 2^2$.

I ended Chapter 1 by pointing out that you couldn't use whole-number exponents to solve all multiplication and division problems, no matter which base you used. With all bases equally imperfect, was there nevertheless one that was superior to the others, regardless of imperfection?

That is where matters stood, and with this superquick recapitulation, we can now go on—

In order to decide among possible bases, let's list the values of some exponential numbers. Let's consider the bases from 2 to 10 and evaluate each for the exponents from 4 to −4, as in Table 3.

If we do this, we find our problem is solved at once. Anyone looking at Table 3 is bound to decide that a series of numbers like 10,000, 1000, 100, 10, and 1 looks neater and easier to handle than a series like 6561, 729, 81, 9, and 1 or any of the others on the table do. The powers of 10 are "round numbers"—that is, they invariably have a lot of round-shaped digits, or 0s. This is not true for any of the others on the list.

If we went up through 11, 12, 13, and so on, we would find that if we used 20, 30, or 40 as a base, we would also get round figures, but in addition there would be digits other than 0s that could create trouble. For instance, $30^5 = 24,300,000$ and $40^4 = 2,560,000$. The advantage of 10 as a base over these others is that when raised to any exponent,

TABLE 3

Exponent	Base→ 2	3	4	5	6	7	8	9	10
+4	16	81	256	625	1296	2401	4096	6561	10,000
+3	8	27	64	125	216	343	512	729	1000
+2	4	9	16	25	36	49	64	81	100
+1	2	3	4	5	6	7	8	9	10
0	1	1	1	1	1	1	1	1	1
−1	1/2	1/3	1/4	1/5	1/6	1/7	1/8	1/9	1/10
−2	1/4	1/9	1/16	1/25	1/36	1/49	1/64	1/81	1/100
−3	1/8	1/27	1/64	1/125	1/216	1/343	1/512	1/729	1/1000
−4	1/16	1/81	1/256	1/625	1/1296	1/2401	1/4096	1/6561	1/10,000

the only digits you will find in addition to all those 0s is a single 1, which is just as easy to handle as 0 is.

The number 10 is not unique in this. If we use 100 or 1000 or any number of that sort, we will still get round numbers made up of 0s plus a single 1. Thus $100^3 = 1,000,000$ and $1000^4 = 1,000,000,000,000$. However, of all numbers of this sort, 10 is the smallest and that represents a great advantage.

Therefore 10 it is. It is 10 that is the common base for logarithms used in ordinary computation. (In calculus, another base is preferable to 10 and that is a number represented by e; it is not an integer but a never-ending decimal with a value equal to 2.71828. . . . But that's another story.)

The usefulness of 10 does not lie in any mystical property possessed by the number, however. It is entirely the result of the fact that we use a 10-based positional number system, which, in turn, is based on the anatomical accident that we happen to have five fingers on each of two hands. Had we evolved with four fingers on each hand, or six, we might well have developed an 8-based system of numeration or a 12-based one, and then 8 or 12 would have served as the most convenient base for calculations.

Suppose we take a closer look at exponential numbers to the base 10, as in Table 4.

As you can see, when 10 is raised to an exponent that is a positive integer, the result is a 1 followed by as many 0s as is the value of the exponent. This is an invariable rule, again based on the nature of our number system and

TABLE 4

10^6	$=$	$10 \times 10 \times 10 \times 10 \times 10 \times 10$	$=$	1,000,000
10^5	$=$	$10 \times 10 \times 10 \times 10 \times 10$	$=$	100,000
10^4	$=$	$10 \times 10 \times 10 \times 10$	$=$	10,000
10^3	$=$	$10 \times 10 \times 10$	$=$	1,000
10^2	$=$	10×10	$=$	100
10^1	$=$	10	$=$	10

nothing else. We can be sure, therefore, that 10^7 is equal to a 1 followed by seven 0s, or 10,000,000—or, in words, ten million. We don't have to go to the trouble of getting the product of seven 10s to be sure of this. In the same way 10^{33} is equal to 1,000,000,000,000,000,000,000,000,000, 000,000 (one billion trillion trillion) and, if you have the patience, you can just as easily, in principle, write the value 10^{2347}. (You can see the advantage of exponential expressions over ordinary numbers.)

Using this system, we have a new way of demonstrating that $10^0 = 1$, something that is obvious on inspection. After all, 10^0 must be equal in value to a 1 followed by zero 0s; that is, to 1.

What about the negative exponents? We can always convert such exponentials into a unit fraction in which the denominator is what the number would be if the exponent were the equivalent positive form. Thus $10^{-4} = 1/10^4 = 1/10,000$ and $10^{-752} = 1/10^{752} = 1/100,000,0 \ldots$ —but you can write that out for yourself.

Fractions are, however, generally clumsy to work with and it is much easier to play with decimals. If we were to convert all the fractions in Table 3 into decimals, we would see another reason why 10 is the best base for computations.

Consider the base 2. In decimals, $2^{-1} = 1/2 = 0.5$; $2^{-2} = 1/4 = 0.25$; $2^{-3} = 1/8 = 0.125$; and $2^{-4} = 1/16 = 0.0625$. Each power has one or more non-0 digits and this introduces increasing complications.

If we consider the negative power of base 3 in decimal form, matters are even worse. Thus $3^{-1} = 1/3 = 0.333333 \ldots$, a never-ending decimal. Then, $3^{-2} = 1/9 = 0.11111 \ldots$; $3^{-3} = 1/27 = 0.0370370370 \ldots$, and $3^{-4} = 1/81 = 0.1234567901234. \ldots$

All the negative powers of 3 are never-ending decimals. (Again, mind you, this is only because we use a 10-based system of numbers. If we used a 9-based system, the negative powers of 3 would be limited decimals and the negative powers of 2 would be unending decimals.)

Consider the values of 10 raised to negative exponents, however. We have $10^{-1} = 1/10 = 0.1$; $10^{-2} = 1/100 = 0.01$;

$10^{-3}=1/1000=0.001$; $10^{-4}=1/10,000=0.0001$; and so on.

All the negative powers of 10, without exception, have values that are (in decimal form) round numbers built up of a number of 0s and a single 1, just as in the case of the positive exponents. The difference is that in the case of the negative exponents the 0s are to the left of the 1 rather than to the right—i.e., before it rather than behind it.

Again, the total number of 0s is equal to the value of the negative exponent, provided you remember, as a matter of convention, to include the one 0 to the left of the decimal point. Thus 10^{-8} is equal to 0.00000001 (eight 0s including the one to the left of the decimal point. Again, 10^{-18} is equal to 0.000000000000000001, while I leave it to you to write the value of 10^{-3098} if you wish to.

Remember, as I said in the last chapter, that the exponents are logarithms. If $10^2=100$, then 2 is the exponent to which 10 must be raised to give 100 and another way of putting it is that the logarithm of 100 is 2. In the same way, since $10^4=10,000$, then the logarithm of 10,000 is 4; and since $10^{-3}=0.001$, then the logarithm of 0.001 is -3.

In shorter form, we can say "log $100=2$," "log $10,000=4$," "log $0.001=-3$," and so on. In a sense, we are just counting the 0s after the 1 for positive logs and before the 1 for negative logs.

Strictly speaking, we should say that "$\log_{10}100=2$"—that is, "the logarithm to the base 10 of 100 is equal to 2." The base 10 is so nearly universally used in ordinary computations, however, that we just leave it out and assume that that's what it is. If, for any reason, we used any base other than 10, *then* we would have to indicate it. (Again e is an exception. If e is used as a base, we speak of "natural logarithms" and use the abbreviation "ln." This is still another story.)

Any computations involving the multiplication or division of powers of ten *only,* become a lot easier if we use logarithms. Thus, if we want to solve the problem $1,000,000 \times 0.001 \div 100 \times 10,000$, we might easily get lost in the 0s.

Suppose instead that we substitute the logarithms, adding them at each multiplication and subtracting at each division.

The logarithm of 1,000,000 is $+6$; that of 0.001 is -3; that of 100 is $+2$ and that of 10,000 is $+4$. The problem is reduced, then, to $(+6)+(-3)-(+2)+(+4)$, or $6-3-2+4$, which is equal to $+5$. The answer, therefore is 100,000.

It looks complicated, if you've never done it before, but if you get used to logarithms, it becomes second nature, and this system of adding and subtracting logarithms can cause you to forget how to do the problem by the ordinary rules of multiplication and division. (And good riddance, too.)

But how often do you deal with powers of 10 only? Suppose you want to deal with the nearly similar problem of $2,300,000 \times 0.0015 \div 120 \times 30,500$. What then?

If we knew the logarithms of each of these numbers we could proceed as before, so let's start with 2,300,000. What is the logarithm of 2,300,000? In other words, 10 must be followed by what exponent to yield 2,300,000?

If we confine ourselves to whole-number exponents, there is no answer, for $10^6 = 1,000,000$ and $10^7 = 10,000,000$ and 2,300,000 is somewhere in between. Or we can say that log $1,000,000 = 6$, log $10,000,000 = 7$, and log 2,300,000 must be some fractional value between 6 and 7.

But how can you have some fractional log? It is equivalent to a fractional exponent, to the raising of 10 to a fractional power. It is easy to understand what 10^6 or 10^7 is, but what the heck is, let us say $10^{6.362}$?

Let's be systematic about it. To begin with, let's write all numbers as much as possible in powers of ten. For instance, 2,300,000 can be written as $230,000 \times 10$, or as 230×1000, or as $0.23 \times 10,000,000$, and so on. There are an infinite number of possibilities, but it is neatest if we agree to have the "non-round" portion of the number— the part that is not a power of 10—fall in the range between 1 and 10. If we follow that rule, we can write 2,300,000 as $2.3 \times 1,000,000$ or as 2.3×10^6.

Anyone for Tens?

The other numbers in the problem cited five paragraphs above can then be rewritten as follows:

$$0.0015 = 1.5 \times 0.001 = 1.5 \times 10^{-3}$$
$$120 = 1.2 \times 100 = 1.2 \times 10^2$$
$$30,500 = 3.05 \times 10,000 = 3.05 \times 10^4$$

Once you get enough practice thinking of numbers in this fashion, it becomes second nature. It actually becomes easier eventually, and makes more sense, to think of a number as 1.2×10^2 than as 120. In fact, to keep the system uniform, there is value in thinking of 100 not as 10^2 but as 1×10^2 and it becomes easier to work with that (believe it or not) than with 100.

The problem we started with above then becomes $2.3 \times 10^6 \times 1.5 \times 10^{-3} \div (1.2 \times 10^2) \times 3.05 \times 10^4$. The exponential part of the problem can be worked out by just adding and subtracting exponents and, at a glance (if you've had practice), you see that that part of the answer is 10^5, so that the problem reduces to $2.3 \times 1.5 \div 1.2 \times 3.05 \times 10^5$.

It doesn't seem to make the problem any easier, because we've just gotten rid of a lot of 0s which aren't really troublesome except for possible mistakes in counting. By taking out the 0s exponentially, we have done what is usually referred to as "locating the decimal point."

Now all we have to do is to work out the logarithms of the numbers between 1 and 10 and we're home free. Unfortunately, that still represents an infinite set of numbers. Even if we succeed in getting the logarithms of 2, 3, 4, 5, 6, 7, 8, and 9 we still have to get all decimals between—numbers such as 2.1 and 3.45 and 8.112 and so on. Unending!

Since $10^0 = 1$ and $10^1 = 10$, all the infinite array of numbers between 1 and 10 must have exponents (and therefore logs) that lie between 0 and 1, and we still have to learn the meaning of a fractional exponent. Suppose we begin with an exponent of 1/2 or, as it can also be written, 0.5. What is the meaning of an expression such as $10^{1/2}$ or $10^{0.5}$?

We know that if we multiply two exponential numbers

with the same base, we must add the exponents; therefore $10^{1/2} \times 10^{1/2} = 10^1 = 10$. This means that $10^{1/2}$ represents some number which, when multiplied by itself, gives us 10. Such a number has long been defined as the "square root" of 10. Thus, 5 is the square root of 25, since $5 \times 5 = 25$; 1.6 is the square root of 2.56 since $1.6 \times 1.6 = 2.56$; and so on.

The square root isn't always an integer or a simple decimal. In fact, it hardly ever is. It is usually an unending decimal, but you don't need all the infinite set of digits in the decimal to have a useful decimal. If you have a few places, you have a square root you can work with and it isn't very difficult to calculate the square root of any number to a few places.

Never mind the details, but the square root of 10 is 3.162. . . . If we take the simple decimal 3.162 as the square root we see that $3.162 \times 3.162 = 9.998244$. That's not exactly 10, but it is close enough for many purposes.

We know, then, that $10^{1/2} = 3.162$ (just about). Conversely, we know that log $3.162 = 0.5$ (just about).

Similarly, $10^{1/3}$ is equal to the cube root of 10 since $10^{1/3} \times 10^{1/3} \times 10^{1/3} = 10^1$. The cube root of 10 can be calculated to be about 2.154; therefore $10^{1/3}$ is just about equal to 2.154 and log $2.154 = 0.333$ (just about).

Again, $10^{2/3}$ is equal to the cube root of 100 since $10^{2/3} \times 10^{2/3} \times 10^{2/3} = 10^2$. Since the cube root of 100 is about equal to 4.642, we know that the log of 4.642 is about equal to 0.667.

We might go on and point out that $10^{1/4}$ is the fourth root of 10 or the square root of $10^{1/2}$ or the square root of 3.162; that $10^{3/4}$ is the fourth root of 1000; and so on.

Mathematicians have other ways of calculating logarithms, but we don't have to be concerned with that. The point I have tried to demonstrate is simply that fractional exponents, and therefore fractional logarithms, *do* have meaning and that, in principle, the logarithm to the base 10 (or to *any* base, for that matter) can be calculated for *any* number.

Logarithms, with inconsiderable exceptions, are, like

square roots, unending decimals, but they can be calculated, in principle, to any number of places, given time. Five places is usually adequate. Thus the logarithms for the integers from 1 to 10 are given in Table 5 to five places; the logarithms for the tenth-numbers between 3 and 4 are given in Table 6; and the logarithms for the hundredth-numbers between 3.2 and 3.3 are given in Table 7. Each of these is an example of a small section of a log table, which I referred to in introductory paragraphs of Chapter 1.

TABLE 5

Number	Logarithm
1	0.00000
2	0.30103
3	0.47712
4	0.60206
5	0.69897
6	0.77815
7	0.84510
8	0.90309
9	0.95424
10	1.00000

TABLE 6

Number	Logarithm
3.0	0.47712
3.1	0.49136
3.2	0.50515
3.3	0.51851
3.4	0.53148
3.5	0.54407
3.6	0.55630
3.7	0.56820
3.8	0.57978
3.9	0.59106
4.0	0.60206

TABLE 7

Number	Logarithm
3.20	0.50515
3.21	0.50651
3.22	0.50786
3.23	0.50920
3.24	0.51055
3.25	0.51188
3.26	0.51322
3.27	0.51455
3.28	0.51587
3.29	0.51720
3.30	0.51851

Now let's return to our problem which was $2.3 \times 1.5 \div 1.2 \times 3.05 \times 10^5$.

Using a log table (a real one—not the small samples I've given you in this chapter), we find that:

log 2.3 = 0.36173
log 1.5 = 0.17609
log 1.2 = 0.07918
log 3.05 = 0.48430

We add and subtract logarithms instead of multiplying and dividing the corresponding numbers, which means we convert the problem to $0.36173 + 0.17609 - 0.07918 + 0.48430$ and that comes out to 0.94294.

We can use the log table to find out what number has the logarithm 0.94294 and the answer turns out to be very nearly 8.769. The answer to our problem is, then, just about 8.769×10^5, or 876,900.

It may seem that using a log table is more trouble than it's worth. First, there's all that looking up of logarithms and copying them down (and making a mistake, possibly, at either step). Then we have to add and subtract numbers in five decimal places which is tedious in itself. Wouldn't it have been simpler to go through the multiplications and divisions instead?

You're welcome to try the multiplications and divisions

in the problem I've given just to see how simple it would be to do so, and please remember that this problem is a rather simple one. If the problem were more complicated and if there were a great many of them to do, you would settle for a log table soon enough. In 1619, when Johannes Kepler was trying to calculate out the orbit of the planet Mars, working with numerous observations and dealing with vast and repetitious multiplications and divisions, his lifetime would not have been enough to carry them all through without error—were it not for the invention of logarithms in 1614.

To avoid the most tedious aspects of logarithms, however, the English mathematician William Oughtred invented the slide rule in 1622. The principle here involves the sliding of one straight-edge against another, each ruled off with numbers spaced carefully and identically. This enables you to add lengths equivalent to the value of the numbers and to attain the sum mechanically (or the difference).

If the numbers from 1 to 10 were spaced equally along each straightedge, and the tenths and hundredths were likewise, you'd be making a straightforward addition or subtraction. What Oughtred did, however, was to space the numbers not equally, but in proportion to the value of their logarithms. For this reason, the space between 1 and 2 is considerably greater than that between 2 and 3, which is in turn greater than that between 3 and 4, and so on. If you look at a slide rule you will see that the higher numbers seem to squeeze together. This reflects the fact that the logarithms increase by smaller amounts as you go up the scale of digits (see Table 5).

If you then slide one straightedge against the other, you are not adding or subtracting lengths equivalent to the values of the numbers, but equivalent to the value of the logarithms of the numbers. You are, in effect, adding or subtracting logarithms and, therefore, multiplying or dividing numbers.

If you use the slide rule constantly, you get very practiced at manipulating it and you can slide the central piece back and forth, solving problems such as $2.3 \times 1.5 \div 1.2 \times 3.05$ in just a few seconds. I'm badly out of practice, and I'm not very deft with my hands even when in practice, but I solved that problem on the slide rule in eighteen seconds and found the answer to be about equal to 8.77. Remembering that I had to multiply this by 10^5, I found the answer to the problem we've been working with in this article to be 877,000.

Now I have two answers: 876,900 by the use of log tables and 877,000 by the use of the slide rule. Which is correct?

Actually, neither, but the log table answer is more nearly correct. The trouble is that with a slide rule of ordinary length (10 inches) it is just possible to get three decimal places, while with the log tables one commonly finds it is easy to get four decimal places.

The slide rule tends to give you a poorer answer, but it gives it to you much more quickly. The tendency, then, is to use a slide rule for quick approximate results and to turn to a log table only when unusual accuracy is desired.

But then along comes a pocket computer which manipulates little flashes of electric current in such a way as to mimic the processes of addition and subtraction. Then, by repeated additions and subtractions in tiny fractions of a second, it performs the processes of multiplication and division. —Even a very cheap pocket computer, such as mine, can do this.

I punch the necessary figures into it and in twelve seconds (I'm not very deft, remember, or it would take less time) find that $2.3 \times 1.5 \div 1.2 \times 3.05 = 8.76875$. Multiplying that by 10^5, we get the answer to the original problem as 876,875, which is exact. The problem was solved more quickly than by the slide rule and more exactly than by a five-place log table, so who needs either?

In fact, we don't even have to take out the exponential parts of the numbers (if we only make sure that we don't put in more 0s than the particular computer is equipped to hold) so that the pocket computer even locates the decimal point for us.

On my computer, I punch 2,300,000, then the multiplication sign, then the 0.0015, then the division sign (partial answer, 3450), then 120, then the multiplication sign (partial answer, 28.75), then 30,500, then the equals sign, and the final answer is 876,875.

Now you know why I had to retire my faithful slide rule after twenty years of exemplary service—and why slide rules aren't manufactured any more.

B ELEMENTS

3. Countdown

There's something inexorable about a countdown.

If we count upward, we can never be certain about the limit, since the series of natural numbers has no upper end. Thus, I have recently celebrated (if that's the word I want) yet another birthday. Naturally, the thought occurs to me, now and then, that there will be some limit to the number of years I will experience, but no matter how high the number reaches, I can always hope for one more.

But after all, there is a "deathday" somewhere in the year, as well as a birthday, even though we don't know what the deathday is in advance (or much in advance, anyway—I admit we might have a sentence of execution to face, either legally or medically). If, every time your deathday came, you knew it as you know your birthday and if you had to subtract one from the years left you, life would be unbearable. A countup can be lived with and the thought of death can be set aside. A countdown is the crack of doom; the approach of zero leaves no room for argument.

That may be why, I think, there are always such difficulties among non-physicists about the speed-of-light limit. Acceleration—going faster and faster—is a countup and any limit seems arbitrary and unacceptable. If you can travel 299,776 kilometers per second, why can't you speed up a little and go 299,777 kilometers per second? If the higher number exists and can be written, why can't it be experienced?

That other scientific ultimate, however—the lower limit

of temperature—is a countdown, so it doesn't upset people. As heat is removed and temperature drops, the amount of heat finally reaches zero and temperature is therefore at an absolute zero, and one can't drop beyond that. When there's no more heat, there's no more heat. Zero is zero and people accept that fact calmly. I've never heard anyone suggest that maybe, if we try a little harder, we can get below absolute zero temperature.

But what gave people the idea that there was an absolute zero temperature in the first place?

To begin with, human beings had to be aware of differences in temperature between sun and shade, between day and night, between winter and summer. Early on, they discovered how to make things artificially warmer by taming fire. Making things artificially colder was more difficult—one way was to rush ice from a mountaintop to a valley and use that ice to keep something temporarily cold despite the surrounding warmth.

Using melting ice as a coolant produces a temperature of 0° Celsius (32° Fahrenheit), but human beings outside the tropics can experience temperatures considerably cooler than that. Temperatures well below freezing are common, and in the northern portions of continental interiors, temperatures as low as −50° C. (−58° F.) are not terribly uncommon.*

But there were no phenomena that human beings associated with cold, until modern times, that seemed to indicate there was some limit beyond which things could grow no colder. In fact, if we leave the world of life out of consideration, there are virtually no visible changes with cold.

Water freezes, to be sure. We all know that. But aside from that one phenomenon, there is almost nothing else. The metal tin turns to another form and crumbles; liquid mercury will eventually freeze; but neither of these changes

---

* In this chapter, I am simply giving temperatures without comment. The question of methods of measuring temperature, especially outside the common range, is for another essay some day.

obtrudes itself on ordinary life. Rocks, ice, air just get colder and colder—why should there be any limit to how cold?

In the last years of the seventeenth century, however, a French physicist, Guillaume Amontons (1663–1705), grew interested in the behavior of gases. He took careful note of the manner in which a column of gas expanded as it grew warmer and contracted as it grew colder.

He was the first to make careful measurements of the changes in gas pressure with temperature. He worked with gas trapped in a container by a column of mercury. When the temperature rose and the gas expanded, Amontons added more mercury to the column, so that the added weight of mercury compressed the gas and restored the original volume. When the temperature fell and the gas contracted, Amontons removed some mercury, reducing its weight and allowing the gas to expand to its original volume. From the height of the mercury column he could measure gas pressures at different temperatures.

By 1699 Amontons discovered in this way that not only did volume grow less as temperature dropped while pressure remained constant, but that pressure did so, too, while volume remained constant. What's more, he found that both the volume and the pressure decreased by a fixed percentage under these conditions for every fixed drop in temperature.

In other words (to use modern terminology), suppose we begin with gas at the freezing point of water, which is 0° C. If we drop the temperature to −1° C.,† then at constant pressure the volume of the gas will decline by $1/x$ of its original amount. If temperature dropped to −2° C., the volume would drop by $2/x$ the original amount. This went on as far down as Amontons could measure.

Amontons could see that if this loss of volume continued at a constant rate all the way down, then by the time the temperature reached a value of $-x°$ C., *all* the volume of

† We can drop below zero in this case because the Celsius zero is an arbitrary one and not an absolute.

a gas would be gone. Since volume can't become less than zero, a temperature of $-x°$ C. would represent an ultimate, an absolute zero beyond which temperature could not drop.

From Amontons's observations, he decided that such an absolute zero would be reached at what we would now call $-240°$ C. ($-400°$ F.).

Amontons's work went unnoticed, but in 1802 the French chemist Joseph Louis Gay-Lussac (1778–1850) conducted similar experiments and made the more nearly accurate estimate that gases lost 1/270 of their 0° C. volume for each Celsius degree below zero. Since then, the estimate has been further improved and the absolute zero of temperature is now taken to be $-273.15°$ C. ($-459.67°$ F.).

There is, of course, a catch to this. Some gases turn to liquids as the temperature drops and then the "gas laws" involving pressure, volume, and temperature, as worked out by men like Amontons and Gay-Lussac, no longer hold. If *all* gases turn to liquids when the temperature drops low enough, then notions of an absolute zero based entirely on the behavior of gases might well prove meaningless. The matter of the liquefaction of gases thus gained theoretical interest.

In 1803 the atomic theory was introduced by the English chemist John Dalton (1766–1844). The matter of liquids and gases made new sense in the light of that theory.

In liquids, the atoms or atom groupings (molecules) are in contact with each other. When a liquid vaporizes, those atoms or molecules separate widely.

If we begin with a gas, cooling causes the atoms or molecules to subside, so to speak, and come closer together. Eventually, they come into contact, at which time they stick together and the gas liquefies. If the gas were put under pressure, that, too, should force the atoms or molecules together and liquefy the gas. If you both lower the temperature *and* increase the pressure, then the liquefaction proceeds even more readily than by either change alone.

The first to use this principle and to go about the process of liquefying gases systematically was the English chemist

Michael Faraday (1791–1867). He did so by following a suggestion made by his mentor, Humphry Davy (1778–1829). (Later on, Davy felt Faraday gave insufficient credit to him for his part in the investigation and that helped give rise to Davy's bitter jealousy of his erstwhile assistant.)

What Faraday did was to use a strong glass tube which he had bent into a boomerang shape. In the closed bottom he placed some substance which, when heated, would liberate the gas he was trying to liquefy. He then sealed the open end. The end with the solid material he placed into hot water. This liberated the gas in greater and greater quantity, and since the gas was in the limited space within the tube, it developed greater and greater pressure.

The other end of the tube Faraday kept in a beaker filled with crushed ice. At that end the gas would be subjected to both high pressure and low temperature and would liquefy. In 1823 Faraday liquefied the gas chlorine in this manner. Chlorine's normal liquefaction point is −34.5° C. (−30.1° F.).

Using this method, Faraday liquefied several other gases, too.

Liquefied gases offered a new means for producing low temperatures.

Suppose a gas has been liquefied under pressure. If the pressure is then decreased, the gas vaporizes. This means that the molecules of the liquid must move apart, and to do this they must gain energy. If the liquid is kept in an insulated container, little of that energy can be gained from the outside world. It must, instead, be gained from the only substance accessible—the liquid itself. In other words, as the liquid evaporates, the temperature of the portion of the liquid that remains unevaporated drops, and further evaporation slows until heat finally reaches the liquid from the outside environment.

In 1835 a French chemist, C. S. A. Thilorier, used the Faraday method to form liquid carbon dioxide under pressure, using metal cylinders which would bear greater pressures than glass tubes. He prepared liquid carbon dioxide in considerable quantity and then allowed it to escape from the tube through a narrow nozzle. The rapid expansion of some of the liquid, evaporating into ordinary gaseous carbon dioxide, reduced the temperature of the remainder so sharply that the carbon dioxide froze. For the first time, solid carbon dioxide was formed.

Liquid carbon dioxide is only stable under pressure. Solid carbon dioxide exposed to ordinary pressures will "sublime"—that is, evaporate directly to gas without melting. The sublimation point is $-78.5°$ C. ($-109.3°$ F.) and solid carbon dioxide is therefore a much more efficient cooling agent than ordinary ice is. Thilorier added solid carbon dioxide to diethyl ether (the common anesthetic, which remains liquid down to very low temperatures). By allowing the mixture to evaporate, he managed to attain temperature as low as $-110°$ C. ($-166°$ F.). The temperature countdown was moving ahead briskly.

In 1845 Faraday returned to the task of liquefying gases under the combined effect of low temperature and high pressure, making use this time of a mixture of carbon dioxide and diethyl ether as his cooling medium. Despite this, and despite the fact that he used higher pressures than before, there were some gases he could not liquefy.

There were six of these in fact: hydrogen, oxygen, nitrogen, carbon monoxide, nitric oxide, and methane. Faraday named them "permanent gases." To the Faraday list, there might be added five more gases which he would not have been able to liquefy had they been known in 1845 (which they were not). These were helium, neon, argon, krypton, and fluorine.

The question as to whether the permanent gases were truly permanent continued to seem theoretically important in 1845. If they could not be liquefied, however low the temperature, then the gas laws applied all the way down and there was an absolute zero. If they could all be liquefied at some temperature lower than had yet been attained, then there was no compelling reason to suppose an absolute zero existed.

* * *

But then, in 1848, the Scottish physicist William Thomson (1824–1907), who, later in life, was given a peerage and became Baron Kelvin, took up the matter from a different standpoint. The science of thermodynamics (involving the interconversions of energy and work) was making great strides and, using thermodynamic principles, Thomson showed there *had* to be an absolute zero.

The energy of motion—"kinetic energy"—of atoms and molecules declines steadily with declining temperature, and, in fact, temperature is only the measure of the average kinetic energy of the atoms and molecules of a given system. The decline in gas volume at constant pressure is only a consequence of this loss of kinetic energy and secondary to it. The rate of loss of kinetic energy with declining temperature is such that at −273.15° C. (−459.67° F.) it is zero. You can then have no lower temperature since you cannot have less than zero kinetic energy.

The importance of this view is that it no longer mattered whether gases liquefied or not. The loss in volume and pressure may become irrelevant once a gas becomes a liquid, but the loss in molecular kinetic energy does not. Hence, we may view Thomson as the true discoverer of the existence of absolute zero.

Thomson also pointed out that thermodynamic calculations were easier and more sensible if the temperature scale was so arranged as to begin at absolute zero. Counting upward in Celsius degrees from absolute zero, we would reach the freezing point of water at a mark of 273.15° in "absolute temperature." The absolute temperature is usually written as "° K.," where the K. stands for "Kelvin." Thus, the freezing point of water (or the melting point of ice) is 273.15° K.

The Scottish engineer William John Macquorn Rankine (1820–72) introduced a variety of the absolute scale in which one counted upward from absolute zero by Fahrenheit degrees. On this scale, the melting point of ice is 459.67° Rank. This scale is hardly ever used—and never by scientists.

* * *

Although the work of Kelvin robbed the liquefaction of gases of some of its theoretical importance, it substituted the lure of the countdown—of setting records for lower temperature and of initiating a race, so to speak, for absolute zero.

The race was given an important intermediate goal as a result of the work of the Irish chemist Thomas Andrews (1813–85), in 1869. In that year, he pointed up the importance of lowering the temperature by showing that pressure was not always useful in producing liquefaction.

Until then, it had been assumed that given enough outside pressure, any gas would have to liquefy at any temperature. After all, if you just squeeze the atoms and molecules close enough to each other, won't they eventually stick together so that the gas will liquefy?

Not so, said Andrews. Working with carbon dioxide, he measured the pressure at which it could be liquefied at various temperatures. As the temperature went up by small amounts the pressure required for liquefaction went up by small amounts, too, as might be expected. At a temperature of 31° C. (88° F.) the orderly progression ceased. Suddenly, even a much greater pressure than had sufficed for liquefaction at a temperature just under that mark, would not make carbon dioxide liquid.

There was a "critical temperature" apparently, above which no pressure would suffice to induce a gas to liquefy. For carbon dioxide, the critical temperature was 31° C. (88° F.), but for other gases, it might be lower.

In 1873, the Dutch physicist Johannes Diderik van der Waals (1837–1923) supplied a theoretical explanation for the concept of critical temperature, and laboratory observations ever since have shown Andrews to have been right.

Now there was an explanation for Faraday's "permanent gases." If the critical temperatures of these gases were below the lowest temperature that could be reached in a bath of solid carbon dioxide and diethyl ether (163° K.), then applying pressure was a waste of effort.

It was no wonder, for instance, that in 1854, the Austrian physicist Johann August Natterer (1821–1901) had failed when he had placed cold oxygen under the then-enormous pressure of 3500 atmospheres. He had not liquefied it and now there was no surprise to it. His cold oxygen was not yet cold enough to be below the critical temperature for oxygen.

As it happens, the critical temperature of four of Faraday's "permanent gases," plus that of four gases not yet isolated even in 1869, is indeed well below the 163° K. mark. The eight are oxygen, argon, fluorine, carbon monoxide, nitrogen, neon, hydrogen, and helium.

In December 1877, the French physicist Louis Paul Cailletet (1832–1913) tackled the problem in the new fashion by concentrating on the cooling. To do this he made use of the "Joule-Thomson effect" which had been worked out by Thomson in 1852 in collaboration with his friend the English physicist James Prescott Joule (1818–89). Joule and Thomson showed that if a gas is allowed to expand, that alone consumes energy as the molecules must move apart and overcome their own small attraction for each other in doing so.

Cailletet began by strongly compressing oxygen, which, of course, grew hot as a result, in a reverse Joule-Thomson effect. He bathed the tube containing this compressed oxygen in cold water to extract as much heat as possible. He then allowed the oxygen to expand very rapidly, using up energy at a far greater rate than could possibly be gained from the outside world. The energy that powered the expansion had to come out of the oxygen's heat, and the temperature of the gas dropped precipitously.

Cailletet managed to lower the temperature of the oxygen to such a degree that it finally liquefied and he obtained a fog of droplets on the wall of the container. Naturally, the fog disappeared quickly as heat leaked inward, but it had been there.

He managed to do the same later on for nitrogen and carbon monoxide, which required even lower temperatures for liquefaction than oxygen did.

Working at the same time as Cailletet, but independently,

the Swiss chemist Raoul Pierre Pictet (1846–1929) also turned the trick. He used a slightly different method, cooling compressed oxygen with liquid carbon dioxide. He managed to get the oxygen at a pressure of several hundred atmospheres and to a temperature of 133° K., a figure which was twenty degrees below oxygen's critical temperature. When he opened an escape valve to the tube containing the oxygen, a jet of liquid oxygen spurted out and, of course, rapidly evaporated.

Both Cailletet and Pictet produced liquid oxygen only momentarily. In 1883, however, two Polish chemists, Karol Stanislav Olszewski (1846–1915) and Zygmunt Florenty Wroblewski (1845–88), used a combination of the methods of Cailletet and Pictet to produce liquid oxygen in quantity. What's more, they managed to cool the oxygen-containing vessels with considerable efficiency by surrounding them with other cold liquids, kept colder by evaporation. As a result, they could for the first time study the properties of the liquid forms of the no-longer-permanent gases at leisure.

Wroblewski died a few years later in a laboratory fire, but Olszewski went on to prepare, in quantity, liquid nitrogen and liquid carbon monoxide.

In 1895 the German chemist Karl von Linde (1842–1934) even made these liquid gases into commercially useful products.

First, he compressed air, cooled it, and allowed it to expand, bringing it to a very low temperature in Cailletet's fashion. Once he had his very-low temperature air, he led it back to bathe a tube that contained compressed air until that compressed air was at the very low temperature he began with. He then let *that* compressed air expand and its temperature dropped from the very low level at which it was to a still lower level.

By repeating this, he was able to produce liquid air at a reasonable price in any quantities desired.

One of the investigators of low temperatures and liquefied gases in the 1890s was the Scottish chemist James

Dewar (1842–1923). Like Wroblewski and Olszewski, he produced liquid oxygen in sizable quantities and studied its properties. He showed, for instance, that liquid oxygen had astonishingly strong magnetic properties and that liquid ozone (a form of oxygen with three atoms per molecule, rather than two) likewise had them.

Dewar gave lectures at the Royal Institution, as once Humphry Davy and Michael Faraday had done (and, for a century, such lectures had been consistently attended by the cream of London society). Naturally, demonstrations involving liquefied gases were startling and impressive, and Dewar took care to give them.

To do it properly, though, he needed to have the liquefied gases easily available, but nevertheless kept in a way that wouldn't allow them to boil away too rapidly.

For that purpose, he devised a special kind of vessel with a double wall. Between the walls was a vacuum.

There are three ways in which heat can be transferred from one point to another point. One is convection, as when matter flows from one point to another, carrying its heat with it. A second is conduction, in which matter stays put but the heat travels along it from one point to another. Neither convection nor conduction is possible across a vacuum.

The third way in which heat can be transferred is by radiation. In this process, heat is converted into photons which leave the heat reservoir altogether and which can then move, be absorbed, and reradiated, at a rate which may be slow but is never zero across any material and even, at the speed of light, through a vacuum.

In the vessel Dewar designed, the side of the inner wall facing the vacuum was mirrored so that the photons were reflected rather than absorbed, thus reducing the radiation effect.

Liquid air placed in such a "Dewar flask" would boil only slowly and would stay liquid for quite a period of time.

Dewar flasks are useful elsewhere than in laboratories. A Dewar flask that can keep liquid air cold can also keep lemonade cold.

When Dewar flasks are used for liquid air, it is better not to stopper them since the air *is* boiling inexorably and the pressure will build up till the stopper is blown out or the flask is shattered. If, on the other hand, you place cold lemonade in the flask, you can stopper it tightly, since no gas is being formed, and then screw on a metal top that can serve as a cup when it is unscrewed.

What you have then is the familiar "Thermos bottle," which can keep your cold liquids cold without the use of ice.

In fact, since the vacuum prevents heat from passing in either way and since the interior of the Thermos bottle is also mirrored, hot liquids will stay hot inside. You can prepare hot coffee, put it in the Thermos, and take off on a picnic, knowing that when you are ready to drink hot coffee, you will have hot coffee—thanks to Dewar.

In twenty years, then, the temperature countdown had dropped from the 163° K. of liquid carbon dioxide and diethyl ether to the 77.3° K. which represented the boiling point of liquid nitrogen. Scientists had gone three quarters of the way down from the familiar and every day freezing point of water to the point of absolute zero.

At the temperature of liquid nitrogen, only three gases are left unliquefied—only three in all the Universe. One was on the list of Faraday's "permanent gases"—hydrogen. In addition, the late 1890s had seen the discovery of the so-called noble gases, and of these, the two lightest, neon and helium, also remained gaseous even at the temperature of liquid nitrogen. They remained to be conquered, and this will be described in the next chapter.

4. Toward Zero

When I was young, I learned the rudiments of English grammar in grade school, and since I quickly learned everything I was taught (either by teachers or by books), I gained the impression that I knew English grammar. Since I hardly ever forget anything I was ever taught, I continued to remain under the impression, as the years passed, that I knew English grammar.

This was a good impression for me to have, for you can imagine how it would have disturbed my sense of inner security if I had known that I was making a living as a writer and *didn't* know English grammar.

But then, four or five years ago, long after I had become established as a writer and had even received praise for the skill and clarity with which I wrote, I picked up a college text on grammar and leafed through it with a condescending air.

That condescension vanished quickly. Not only did it turn out, almost at once, that I knew scarcely anything about English grammar above the grade-school level—but I didn't even know the terminology. I closed the book a shaken man and from that day to this I've never had the nerve to argue with a copy editor.

Of course, I retain my grade-school level of knowledge. I know some simple rules, such as the one that says you can't use comparatives or superlatives for adjectives that represent absolute qualities.

For instance, something that is unique stands absolutely alone. Nothing, therefore, can be "more unique" than something else and you can't speak of anything as the "most unique" of its class. Similarly, since "perfect" implies no flaws at all, you can't take something that is "perfect" and talk of making it "more perfect," since you can't have less than no flaws.

You want to bet on that? Let's see—

In Chapter 3 I discussed the liquefaction of gases and the attaining of low temperatures. By the end of the chapter, we were in the early 1890s and oxygen and nitrogen had been liquefied. Temperatures of about −200° C., or about 70 degrees above absolute zero (70° K.), had been attained.

Oxygen and nitrogen had been liquefied by means of the Joule-Thomson effect. This describes what happens when a gas is allowed to expand. When this happens, the molecules, in moving apart, must overcome the tiny attractions between them. It takes energy to do this and the molecules must obtain the energy from somewhere. The most immediately available energy is that of their own heat so the temperature of the expanding gas drops.

Another way of looking at it is this. The molecules of gas at a particular temperature are moving at some average speed. If the gas is allowed to expand and the molecules move apart, the slight attractions between them pull at the molecules and slow down that speed. Since the average speed of moving molecules is a measure of the temperature, the slowing of the speed means that the temperature drops.

The expanded gas, which is colder than it was when it started, can now be used to cool off a second sample of gas that is still unexpanded. If this cooled second sample is now allowed to expand, its temperature drops lower still. This still colder gas is used to cool an unexpanded sample which is allowed to expand—and so on. Eventually, the gas is cooled to the point of liquefaction.

But the Joule-Thomson effect works only because there is that slight attraction between the molecules. For different gases, there is a different degree of attraction, and the smaller the attraction, the lower the liquefaction point. After all, a gas only turns to liquid because once the molecules move slowly enough, they lack the energy to overcome the attraction, and therefore collapse into contact. The lower the intermolecular attraction, the more slowly the molecules must move (the lower the temperature) before they fail to overcome the attraction and liquefy.

If there were no attraction at all between the molecules, then a gas would never liquefy but would remain a gas even at absolute zero. Such a substance would then be a "perfect gas" or an "ideal gas."

In the case of a perfect gas, expansion would not have to overcome the mutual attraction of molecules, since there was none, so that no energy would have to be withdrawn from the gas and the temperature would not fall. The temperature change on expansion would be zero and the Joule-Thomson effect would therefore not exist in a perfect gas.

Chlorine at ordinary temperatures is not a very close approach to a perfect gas; fluorine is better; oxygen still better; nitrogen even better. You would expect that hydrogen, which remains gaseous even at temperatures of liquid nitrogen, would have a smaller intermolecular attraction than any of the liquefied gases and would represent a still better approach to the perfect gas than any of them.

But it would still only be an approach. Hydrogen would be a nearly perfect gas, but surely never a quite perfect one. The Joule-Thomson effect would be smaller for hydrogen, but we would still presume it to be there. Though cooling through expansion would be a more tedious process for hydrogen than for oxygen or nitrogen, surely it would work eventually and end by liquefying hydrogen.

Not so! When hydrogen was allowed to expand under conditions that would have cooled oxygen or nitrogen, the Joule-Thomson effect did not work. It was not merely that hydrogen did not cool on expansion, as might be the case if it were a perfect gas; hydrogen actually *warmed* on expansion. The Joule-Thomson effect went into reverse!

As a result, some chemists began to speak of hydrogen as a "more-than-perfect" gas.

There you are, grammarians! Chemists know exactly what they mean by a "perfect gas," but hydrogen is more perfect; or, if you prefer, hydrogen is "perfecter."

But how do we explain this?

Since about 1800 the relationship between the pressure, volume, and temperature of a gas could be expressed by means of a very simple equation, called an "equation of state."

If a gas is well above its liquefaction point, the equation of state expresses the properties of the gas almost exactly. The lower the temperature, the less exact the expression is. In this sense, you could define a perfect gas as one for which the equation of state is the exact expression of its properties at all temperatures. By this definition, no real gas is quite perfect, not even hydrogen.

In 1873 the Dutch physicist Van der Waals, whom we met in Chapter 3, was the first to modify the equation of state in such a way as to make it apply reasonably well to real gases at all temperatures.

He suggested that the source of the imperfection lay in the fact that real gases had a small intermolecular attraction and that, in addition, the gas molecules had a definite, albeit small, size. In a perfect gas, the intermolecular attraction would be zero and the volume of the molecules would be zero.

Van der Waals introduced two constants, a and b, into the equation of state. The first, a, was related to the intermolecular attraction and the second, b, to the molecular volume. For each gas, a and b had definite values characteristic of that gas.

If the equation of state were modified to include a and b, it would describe the properties of a particular gas much better than the original "perfect gas" equation of state would.

The equation that describes the Joule-Thomson effect, as deduced from Van der Waals's equation of state, includes the expression

$$\frac{2a}{RT} - b$$

In this expression, a and b are the Van der Waals constants, which are different for each gas; R is the "gas constant," which is the same for all gases; and T is the absolute temperature of the system.

As long as the value of $2a/RT$ is greater than b, then $2a/RT-b$ is a positive number and there is a Joule-Thomson effect. The smaller the positive number, the smaller the Joule-Thomson effect.

In a perfect gas, a and b would both be equal to zero and in that case $2a/RT-b$ would be $0-0$, or 0. For a perfect gas, there would be no Joule-Thomson effect.

Since the value of b is very small in a real gas, always smaller than a and usually much smaller, it is not surprising that $2a/RT$ is usually larger than b and that the value of the expression is positive.

For any given gas, though, the values of a, b, and R are constant and do not change. The value of T, however, represents the temperature and that is easily changed. If we warm any gas, the value of T goes up. Since T is in the denominator of a fraction, its increasing value means that the value of $2a/RT$ goes down. (Consider the fractions $1/2$, $1/4$, $1/8$. . . .)

As the temperature goes up and the value of $2a/RT$ goes down, a point is reached, finally, where $2a/RT$ becomes smaller than b. The expression $2a/RT-b$ then becomes negative. The Joule-Thomson effect does not merely cease; it goes into reverse.

For every gas, there is some temperature, the "inversion temperature," above which the Joule-Thomson effect goes into reverse.

That does not, however, mean that the gas is more than perfect. Whether the value of $2a/RT-b$ is positive or negative doesn't matter; that value is pertinent only if a and b both have certain positive values and the gas is therefore not perfect. If the gas were perfect, a and b would both be zero and the expression would never be either positive or negative, but would always work out to zero. —Even at the inversion temperature, where the value of the expression *is* zero, that is merely because one aspect of imperfection just happens to cancel the other aspect.

So the grammarian is right after all!

For most gases, the value of a is in the neighborhood of a hundred times that of b and the temperature must rise pretty high to lower the value of $2a/RT$ to the point where it is equal to b. For oxygen, the inversion temperature is at 1058° K.

This is an enormously high temperature by ordinary standards and certainly no one attempting to liquefy oxygen would ever begin with oxygen any warmer than room temperature, which is usually just under 300° K.

As gases approach the perfect, the values of a and b both drop, but a drops the faster. Thus, for hydrogen, the value of a is only nine times greater than b. This means that the inversion temperature has got to be far lower for hydrogen than for oxygen.

For hydrogen, the inversion temperature is 190° K. The Joule-Thomson effect is in reverse for hydrogen, therefore, whenever its temperature is higher than it would be in an Antarctica winter at its coldest.

Before hydrogen can be cooled down by the Joule-Thomson effect, then, it must first be cooled down in some other way to get it below its inversion temperature. James Dewar, to whom I referred toward the end of Chapter 3, realized this and used liquid nitrogen for the preliminary cooling of hydrogen gas.

Once hydrogen was at liquid nitrogen temperatures, it was well below its inversion temperature and the Joule-Thomson effect could be used to cool it further. In 1895 Dewar finally obtained liquid hydrogen in quantity.

Hydrogen has a liquefaction point of 20.3° K. When liquid hydrogen is allowed to evaporate, its molecules rush apart into vapor and the energy required for that is withdrawn from what remains of the liquid. The temperature of the liquid hydrogen drops and part of it solidifies while part of it evaporates. The solidification point of hydrogen is 14.0° K.

At liquid hydrogen temperatures—14.0° K. to 20.3° K.—almost everything has become solid. Nitrogen solidifies at 63.3° K. and oxygen at 54.7° K. Even neon which, like hydrogen, is gaseous at liquid nitrogen temperatures, liquefies at 27.2° K., and solidifies at 24.5° K.

Only one substance, helium, remains a gas at liquid hydrogen temperatures. Dewar failed to liquefy it, and it remained the one unconquered gas as the twentieth century opened.

Helium was tackled by a Dutch physicist, Heike Kamerlingh-Onnes (1853–1926), who established the first elaborately equipped laboratory to be devoted entirely to low-temperature work. Helium was even more nearly perfect than hydrogen and, for it, *a* was only 1.5 times as large as *b* so that its inversion temperature was even lower than that of hydrogen.

Kamerlingh-Onnes used liquid hydrogen itself for a preliminary cooling of helium and brought its temperature low enough for the Joule-Thomson effect to take over. In 1908 Kamerlingh-Onnes liquefied helium at a temperature of 4.2° K.

He then allowed the liquid helium to evaporate in an attempt to lower its temperature still further and obtain solid helium, but he didn't succeed. He managed to get the helium temperature down to 0.83° K. before he died, but it remained stubbornly liquid.

In 1926, though, a few months after Kamerlingh-Onnes died, his co-worker, William Keesom, applied pressure to very cold liquid helium and managed to solidify it at last.

As it happens, liquid helium, at ordinary pressures, stays liquid right down to absolute zero. The uncertainty principle requires that atoms and molecules retain some residual energy of motion even at absolute zero, and so small is the intermolecular attraction of helium atoms that even this irreducible residual energy is enough to keep helium from solidifying. At temperatures below 1.0° K., however, pressure will do the trick.

* * *

With helium liquefied and solidified, the game would seem to be over. There would seem to be no purpose to be gained in getting rid of that final degree of temperature. In fact, there would seem to be not only no purpose, but no possibility, either.

In 1906 the German chemist Walther Hermann Nernst (1864–1941), worked out what is called the "third law of thermodynamics." From that third law one can deduce that halving the absolute temperature always takes the same effort regardless of the starting point.

If you start at 4° K., for instance, you can with a certain effort reach 2° K. An equivalent effort will next bring you to 1° K.; then to 1/2° K.; then to 1/4° K.; then to 1/8° K.; and so on.

You will keep approaching the absolute zero more and more closely, but the road ahead will remain just as long, in terms of effort, as it ever was, no matter how close you get in terms of temperature figures. It will take an infinite effort to reach absolute zero—which means that in real terms, you can never reach it.

And, indeed, when Kamerlingh-Onnes and Keesom were trying to get lower and lower temperatures in order to solidify helium, it was like slogging through hardening cement. The advance was slower and slower and stalled at about 0.4° K.

Yet scientists couldn't quit.

In 1911 Kamerlingh-Onnes had been measuring the electrical resistance of mercury at lower and lower temperatures. He was quite certain that the resistance would get lower and lower and approach zero as he approached a temperature of absolute zero. It would be nice, though, to have actual observations of the fact.

But then, at 4.12° K., a temperature well above absolute zero, the resistance of the mercury dropped to nothing —not just to nearly nothing but to an actual *zero* as nearly as our best measurements can tell us. An electrical current which is initiated in a ring of mercury at temperatures below 4.12° K. continues indefinitely without any sign of diminution.

The phenomenon is called "superconductivity" and it has been found in a variety of other metals and alloys, each with a different critical temperature below which it becomes superconductive. A few alloys have been found to become superconductive at temperatures of over 20° K., or just within the liquid hydrogen range.

The phenomenon was so unusual that scientists were galvanized into activity. It had to be further studied.

It is now quite clear that understanding superconductivity, as well as other peculiar phenomena observed in the neighborhood of absolute zero, requires the subtle application of quantum theory. It is only at very low temperatures, when the random, jittery, every-which-way motion of atoms that we refer to as "heat" is suppressed, that the quantum theory effects can make themselves felt; and a study of these effects can then give us some very basic understanding of how matter behaves.

(To be crass about it, superconductivity can also have some very important practical uses in terms of transporting electrical power, setting up ultrastrong magnets, building ultraefficient computers, and so on, and so on, and so on.)

After 1911, therefore, the search for ever lower temperatures became not just a record-setting advance, but a push forward to study odd phenomena that could not be approached in any other way.

The technique of allowing gases to expand and liquids to evaporate had reached dead end in the 1920s at about half a degree removed from absolute zero. Something else was needed.

In 1926 the Dutch chemist Peter Joseph Wilhelm Debye (1884–1966) and the American chemist William Francis Giauque (1895–   ), independently suggested a new technique.

The idea involved the use of certain paramagnetic salts, such as gadolinium sulfate. In such salts, the metal atoms act like tiny magnets. In the presence of a strong magnetic field, all the atoms line up in one direction and the salt is magnetized. If the magnetic field is removed, the atoms jiggle around randomly and the salt loses its magnetic properties.

In magnetizing the salt, work is done on it and its temperature rises. This is analogous to the way in which the temperature of a gas goes up when you compress it.

Contrariwise, when the magnetized salt loses its magnetism, the atoms do work pulling apart from each other and gain the energy to do so from their own heat content so that the temperature falls. This is analogous to the way in which the temperature of a gas goes down when you let it expand.

Instead of the usual method of refrigerating by alternately compressing and expanding a gas, always cooling the compressed gas before you allow it to expand, you could perform the analogous task of magnetizing and demagnetizing a paramagnetic salt, always cooling the magnetized salt before allowing it to demagnetize.

If you magnetize such a salt and then cool it in evaporating liquid helium so that its temperature, while magnetized, is brought to 0.8° K. and *then* allow it to demagnetize, the temperature drops precipitously. (To be sure, it will only drop a few tenths of a degree, since that's all there's room for, but such a drop at such a starting temperature is equivalent to an enormous drop at ordinary temperatures—so states the third law of thermodynamics.)

It wasn't till 1933 that Giauque could get the system to working properly. In that year, he used gadolinium sulfate to produce a temperature of 0.25° K. That same year, Dutch chemists using cerium fluoride obtained a temperature of 0.13° K. and then, later in the year, using cerium ethyl sulfate, a temperature of 0.0185° K.

By 1933, then, scientists were suddenly within one twelfth of a degree of absolute zero. Since then, the use of the magnetization technique has brought temperatures as low as 0.003° K.

So far, I have been talking about helium as though it

were *a* substance. It isn't. It is a mixture of two substances. One is helium-4, with an atomic nucleus made up of two protons and two neutrons, and the other is helium-3, with an atomic nucleus made up of two protons and one neutron.

The two substances are by no means identical. For instance, in the gaseous state, helium-3 is only three fourths as dense as helium-4 is. At low temperatures there are some very important additional differences.

These low-temperature differences were not immediately apparent to early experimenters, since helium-3 is so uncommon and thus very hard to work with in reasonable quantity. Helium itself is not a very common substance to begin with and in nature only one helium atom out of a million is helium-3.

It was not till after World War II that helium-3 began to be studied as a substance in its own right, and not till 1949 that it was liquefied. It turned out that helium-3 has a lower liquefaction point than helium-4 does and that it holds the absolute record in this respect.

At a temperature of 4.2° K., when helium (actually helium-4) liquefies, helium-3, if it could be purified in perceptible quantities, would be seen to be still a gas even at that low point on the temperature scale. The liquefaction point of helium-3 is 3.2° K.

In 1928 Keesom had discovered that at 2.2° K. helium-4 changed from one liquid of normal properties (helium I), essentially like those of liquids generally, into another kind of liquid (helium II) that existed only at temperatures below 2.2° K. and had properties that were utterly different from anything observed in any other substance.

Helium II was, for instance, "superfluid" and could move through very small orifices without any measurable friction. A substance could be gas-tight without being helium II-tight. By 1938 the Soviet scientist Peter Leonidovich Kapitza (1894–) was studying such helium II properties in detail.

Helium-3 did not show a helium II phase at any temperature that could be reached, and there was an attempt to get closer and closer to absolute zero to obtain a helium II version of helium-3. Its presence or absence would affect the theories being worked out for low-temperature behavior.

In 1956 the American physicists William Martin Fairbank (1917–   ) and Geoffrey King Walters (1931–   ) discovered that helium-3 and helium-4 did not mix with each other freely at all temperatures. At temperatures below 0.88° K., a mixture of the two separated into two liquids, one of which was high in helium-3 and one of which was low.

In 1962 the German-English physicist Heinz London (1907–   ) suggested that these separated liquids be used as a refrigeration device. If the two liquids are in contact and if helium-3 is pumped away from the liquid which is already low in helium-3 (7 percent helium-3 and 94 percent helium-4), that helium-3 is replaced by an influx from the high (almost 100 percent) helium-3 liquid.

The helium-3 atoms moving out of the high helium-3 liquid are predominantly the fastest moving ones. Those that remain behind have a lower average speed and this is equivalent to a drop in the temperature.

The helium-3 pumped off from the low helium-3 liquid can be cooled and then added to the high helium-3 liquid so that the process is made continuous. Using this helium-3 method, temperature can be reduced to at least as low a mark as those produced by magnetization but with a further advantage—

By magnetization, very low temperatures could only be maintained for a couple of hours; but by use of helium-3, very low temperatures could be maintained for weeks.

In 1965 the helium-3 method was producing temperatures of 0.20° K., and since then a combination of the helium-3 and magnetization methods has produced temperatures as low as 0.00002° K.—temperatures within 1/50,000 of a degree of absolute zero.

And in 1972 it was found that helium-3 *did* shift to a helium II liquid form at temperatures below 0.0025° K.

C EARTH

5. The Floating Crystal Palace

Last month (as I write this) my wife, Janet, and I crossed the Atlantic on the *Queen Elizabeth 2;* then, after one day in Southampton, we crossed right back.

We did this for a number of reasons. I have a pair of talks each way, Janet is crazy about ships, and both of us found ourselves in an island of peace away from the cares of the workaday world. (Actually I managed to write a small book while on board, but that's another story.)

In one respect, though, I was disenchanted on this particular voyage. It had always been my dim assumption that there was one word that is absolutely taboo on any liner. You might say something was "very large," "huge," "monstrous," "gigantic," but you would *never* say something was—well, the adjective begins with a "t."

I was wrong. One evening on the ship, a stand-up comedian said, "I hope you'll all be joining us at the big banquet tomorrow, folks. We're celebrating the anniversary of the *Titanic.*"

I was shocked! Heaven knows I've never been accused of good taste in my off-hand humor, but this, I thought, was going too far. Had I known he was going to say it, I might well have tried to round up a committee for the feeding of poor, deserving sharks by throwing the comedian overboard.

Did others feel the same way?

No, sir! The remark was greeted with general laughter, with myself (as far as I could tell) the only abstainer.

Why did they laugh? I thought about it, and an essay began to build itself in my mind. Here it is—

Let's start with St. Brendan, an Irish monk of the sixth century.

At that time, Ireland could fairly lay claim to being the cultural leader of the Western World. The West European provinces of the Roman Empire lay sunk and broken in gathering darkness, but the light of learning burned in Ireland (which had never been part of the Empire) and the knowledge of Greek was retained there, though nowhere else in the West. Until the Irish light was extinguished by the Viking invasions of the ninth century and the English incursions thereafter, there was a three-century golden age on the island.

Part of the golden age was a set of remarkable Irish explorations that reached to Iceland and perhaps even beyond. (An Irish colony may have existed on Iceland for a century, but was gone by the time the Vikings landed there in the ninth century.) One explorer we know by name was St. Brendan.

About 550, St. Brendan sailed northward from the west coast of Ireland and seems to have explored the islands off the northern Scottish coast—the Hebrides, the Orkneys, the Shetlands. It is possible that he went still farther north, reaching the Faeroe Islands, about 750 kilometers (470 miles) north of the northern tip of Ireland. This was almost surely the record northward penetration by sea by any human being up to that time.

St. Brendan's voyage was remarkable enough for its time, but in later years tradition magnified it. In 800 a fictional account of his voyages was written and proved very popular. It was, in a way, a primitive example of science fiction, in that the writer drew liberally on his imagination but made careful use of traveler's tales as the supporting framework (just as modern science fiction writers would use scientific theory for the same purpose).

In the tale, for instance, St. Brendan is described as having sighted a "floating crystal palace."

Is there anything in oceanic exploration that could give rise to this particular fantasy?

Certainly. An iceberg. Assuming this interpretation to be correct, this is the first mention of an iceberg in world literature.

In later centuries when the northern ocean was systematically explored, icebergs came to be a common sight. Where did they come from?

To be sure, the sea tends to freeze near the poles, and the Arctic Ocean is covered with a more or less unbroken layer of ice in the winter months. This sea ice is not very thick, however. The average thickness is 1.5 meters (5 feet) and some parts may reach a thickness of as much as 4 meters (13 feet).

We can imagine pieces of that sea ice breaking off as the weather warms in the spring and then floating southward, but those pieces would scarcely be impressive. They would be flat slabs of ice, topping sea level by some 40 centimeters (15 inches) or less.

Compare this to an Arctic iceberg, the top of which can tower 30 meters (100 feet) above sea level. One iceberg has been reported as having a record height of 170 meters (560 feet) above sea level—almost half as tall as the Empire State Building. Counting the portion that was submerged, that piece of ice may have been 1.6 kilometers (1 mile) from top to bottom.

Such a huge chunk of ice could have been spawned only on land.

At sea, the liquid water below the ice layer acts as a heat sink which, even in the coldest polar winter, keeps the ice from growing too thick. On land, the solid surface, with less heat capacity than water and with no currents to bring warmer material from elsewhere, drops to low, subfreezing temperatures and exerts no melting effect. The snow simply piles up from year to year and is capable of forming great thicknesses of ice.

Long-lived ice forms and thickens on mountain heights all over the world. It also forms at sea level in polar regions. The largest piece of land in the Arctic which is wholly polar is Greenland and it is on that vast island that the ice is most extensive and thickest.

The Greenland ice sheet fills the interior of the island and is about 2500 kilometers (1500 miles) long, north to south, and up to 1100 kilometers (700 miles) wide, east to west.

The area of the Greenland ice sheet is just over 1,800,000 square kilometers (700,000 square miles)—a single piece of ice, in other words, that is about 2.6 times the area of Texas. At its thickest point, the Greenland ice sheet is about 3.3 kilometers (2 miles) thick. Along most of the Greenland coast, however, is a fringe of bare land that, in places, is up to 300 kilometers (190 miles) wide.

(It is Greenland's southwestern fringe of bare land to which Viking colonists stubbornly clung for four centuries, from 980 to 1380.)

Each year more snow falls on the Greenland ice sheet and hardly any of it melts in the warmer months (and what does, tends to refreeze the next winter) yet the ice sheet does not get endlessly thicker. Ice, you see, is plastic under pressure.

As the ice sheet thickens, its own weight tends to flatten it and spread it out. The ice, driven by enormous pressure, is forced, in the form of glaciers, to move, like solid creeping rivers, along the valleys and into the seas. These Greenland glaciers move at rates of up to 45 meters (150 feet) per day, which is an enormous speed when compared to the rate at which ordinary mountain glaciers (driven by far smaller pressures) move.

When the Greenland glaciers reach the sea, the ice does not melt appreciably. Neither the Greenland sun nor the cold seas surrounding Greenland will deliver enough heat to do much to them. The tips of the graciers simply break off ("calve") and huge lumps of ice plop into the sea. It is these that are the icebergs. (*"Berg,"* by the way, is German for "mountain.")

In Arctic waters, some 16,000 icebergs are calved each year. About 90 percent of them originate from Greenland glaciers that enter the sea in Baffin Bay, which bathes the western shore of the island.

The largest glacier in the world, the Humboldt Glacier, lies in northwestern Greenland at latitude 80° N. It is 80 kilometers (50 miles) across its coastal foot, but it is too cold to break off icebergs at a record rate. Farther south, about two thirds of the length down the western coast of Greenland, the Jakobshavn Glacier calves 1400 icebergs a year.

Since ice has a density of 0.9, most of any iceberg is below the surface. The exact quantity submerged depends on how pure the ice is. The ice usually contains a great many air bubbles which give it a milky appearance rather than the transparency of true ice, and this lowers its density. On the other hand, in approaching the sea, the glaciers may well scrape up gravel and rock which may remain with the iceberg and which would increase its overall density. On the whole, anything from 80 to 90 percent of the iceberg is submerged.

As long as icebergs remain in Arctic waters, they persist without much change. The freezing water of the Arctic Ocean will not melt them appreciably. The icebergs that form off the western coast of Greenland linger in Baffin Bay for a long time, but eventually begin to move southward through Davis Strait into the waters south of Greenland and east of Labrador.

Many icebergs are trapped along the bleak coast of Labrador and there they break up and slowly melt, but some persist, largely intact, as far south as Newfoundland, taking up to three years to make the 3000-kilometer (1800-mile) journey.

Once an iceberg reaches Newfoundland, however, its fate is sealed. It drifts past that island into the warm waters of the Gulf Stream.

In an average year some 400 icebergs pass Newfoundland and move into the shipping lanes of the North Atlantic. Most of them melt in two weeks in the warm embrace of the Gulf Stream, but the remnants of one giant were sighted on June 2, 1934, at the record southerly latitude of 30° N., the latitude of northern Florida.

At the start of the last stage of its journey, however, an iceberg is still massive and menacing and is even more dangerous than it looks, since the major portion of it is submerged and may jut outward considerably closer to some approaching vessel than the visible upper portion does.

In the years before radio, when ships were truly isolated and there was no way of knowing what lay beyond the horizon, icebergs were dangerous indeed. Between 1870 and 1890, for instance, fourteen ships were sunk and forty damaged by collision with icebergs.

Then came the *Titanic*. The *Titanic* was, when it was launched in 1911, the largest ship in the world. It was 270 meters (883 feet) long and had a gross tonnage of 46,000 metric tons. Its hull was divided into sixteen watertight compartments and four of them could be ripped open without sinking the ship. In fact, the ship was considered unsinkable and was proclaimed as such. In April 1912 it set off on its maiden voyage from Southampton to New York, carrying a glittering load of the rich and the socially prominent.

On the night of April 14–15, it sighted an iceberg at a point some 500 kilometers (300 miles) southeast of Newfoundland. The ship had been ignoring the possibility of icebergs and was going far too fast in its eagerness to set a world record for time of crossing. Consequently, by the time the iceberg was sighted, it was too late to avoid a collision.

The collision, when it came, opened a 90-meter (300-foot) gash on the ship's starboard side. A fatal five compartments were sliced open and even so the *Titanic* held out gamely. It took nearly three hours for it to sink.

That might have been enough to save the passengers, but there had been no lifeboat drills and, even if there had been, the lifeboats available had room for less than half the more than 2200 people aboard.

By now, radio was in use on ships and the *Titanic* sent out a distress signal. Another ship, the *Californian*, was equipped to receive the signal and was close enough all night to speed to the rescue, but it had only one radio operator and a man has to sleep sometime. There was no one on duty when the signal came in.

More than 1500 lives were lost when the *Titanic* went down. Because of the drama of that sinking, the number of lives lost, and the social position of many of the dead, the disaster revolutionized the rules governing sea travel. After the tragedy, all passenger ships were required to carry lifeboats with enough seats for everyone on board, lifeboat drills were to take place on every passage, radio receivers were operating twenty-four hours a day, with men taking shifts at the earphones, and so on.

In addition, in 1914 an International Ice Patrol was established and has been maintained ever since, to keep watch over the positions of these inanimate giants of the deep. It is supported by nineteen nations and is operated by the United States Coast Guard. The patrol supplies continuing information on all icebergs sighted below latitude 52° N., with a prediction of the movements of each over the next twelve hours.

Eventually, air surveillance and radar were added to the patrol and in the years since it has been established not one ship has been sunk by an iceberg within the area under guard. Indeed, modern liners stay so far away from icebergs that passengers never even see them on the horizon. It's no wonder, then, that the passengers on the *Queen Elizabeth 2* could afford to laugh at a tasteless reference to the *Titanic*.

The glaciers of western Greenland are the most dangerous iceberg formers in the world, but not the largest. They can't very well be the largest since the Greenland ice sheet, while the second largest in the world, is a very poor second.

The largest ice sheet is that of Antarctica. The Antarctica ice sheet is a roughly circular mass of ice with a diameter of about 4500 kilometers (2800 miles) and a shoreline of over 20,000 kilometers (12,500 miles). It has an area of about 14,000,000 square kilometers (5,500,000 square miles) and is about 7 1/2 times the Greenland ice sheet in area—and about 1 1/2 times the area of the United States. The average thickness of the Antarctica ice sheet is just about 2 kilometers (1 1/4 miles), and at its thickest, it is 4.3 kilometers (2 2/3 miles).

The total volume of the Antarctica ice sheet is about 30,000,000 cubic kilometers (7,000,000 cubic miles) and this is 90 percent of all the ice in the world.

There are two deep indentations into the roughly circular continent, and these are Ross Sea and Weddell Sea. As the Antarctica ice sheet is flattened out and spreads outward, it reaches these seas first, but it doesn't calve there as the western Greenland ice sheet does. The Antarctica ice sheet is too thick and, instead, it moves out intact over the seas to form two ice shelves.

The ice shelves remain intact for a distance of up to 1300 kilometers (800 miles) out to sea and form slabs of ice that are some 800 meters (1/2 mile) thick where they leave land and are still 250 meters (1/6 mile) thick at their seaward edge. The Ross Ice Shelf, the larger of the two, has an area equal to that of France.

The ice shelves do not push northward indefinitely, of course. Eventually, slabs of ice break off the seaward edge, to form huge "tabular icebergs," flat on top, up to 100 meters (330 feet) above sea level, and with lengths that can be measured in hundreds of kilometers.

In 1956 a tabular iceberg was sighted that was 330 kilometers (200 miles) long and 100 kilometers (60 miles) wide—a single piece of free-floating ice with an area half again that of the state of Massachusetts.

For the most part, Antarctic icebergs drift in the Antarctic Ocean and are carried round and round Antarctica, edging northward and slowly melting. Although representing a much larger mass of ice in total than the 400 Greenland icebergs that slip past Newfoundland each year, the Antarctic icebergs scarcely impinge upon the consciousness of mankind, since they are well away from the chief ocean trade routes of the world. Nowhere in the Southern Hemisphere are there shipping lanes as crowded as those of the North Atlantic.

An occasional Antarctic iceberg drifts quite far north-ward; in 1894 the last remnant of one was sighted in the western South Atlantic at latitude 26° S., not far south of Rio de Janeiro, Brazil.

Icebergs are not all bad. The vast Antarctica ice sheet and the huge icebergs it spawns serve as air-conditioners for the world and, by keeping the ocean depths cold, allow sea life to flourish.

Anything else? Well, let's start at another point.

The average American, drinking eight glasses of water a day, will consume 0.7 cubic meters (180 gallons) in a year. There is also water required for bathing, washing the dishes, watering the lawn, and so on, so the average American consumes, at home, 200 cubic meters (53,000 gallons) of water per year.

But Americans also need water for domestic animals, for growing crops, and for industry. To make a kilogram of steel requires 200 kilograms of water, for instance, and to grow a kilogram of wheat requires 8000 kilograms of water.

All told, the water use of the United States comes to 2700 cubic meters (710,000 gallons) per year per person.

In those regions of the world where industry is negligible and where agricultural methods are simple, water needs can be satisfied by 900 cubic meters (240,000 gallons) per person per year. The average figure for the world as a whole might come to 1500 cubic meters (400,000 gallons) per person per year.

How does this compare with the water supply of the world?

If all the water in the world were divided equally among the 4 billion people on Earth right now, it would amount to 320,000,000 cubic meters (85 billion gallons) for each person. That sounds like plenty. This is enough water, if efficiently recycled, to supply the needs of 210,000 times the present world population.

But wait! Fully 97.4 percent of all the water on Earth is the salt water of the ocean, and human beings don't use salt water, either for drinking, washing, agriculture, or industry. That 1500 cubic meters per person per year refers to fresh water only.

If all the *fresh* water on Earth were divided equally among the 4 billion people on Earth right now, it would amount to 8,300,000 cubic meters (2.2 billion gallons) for each person. Still not terrible. With efficient recycling, the fresh water supply could support 5500 times the present world population.

But wait! Fully 98 percent of all the fresh water on Earth is locked up in the form of ice (mostly in the Antarctica ice sheet) and it isn't available for use by human beings. The only water that human beings can use is *liquid* fresh water, found in rivers, ponds, lakes, and ground water and replenished continually by rain and melting snow.

If all the *liquid* fresh water were divided equally among the 4 billion people on Earth right now, it would amount to 160,000 cubic meters (42,000,000 gallons) per person per year. That's still not fatal. With efficient recycling, that is enough to support 100 times the present population on Earth.

But wait! The recycling isn't 100 percent efficient. We can't very well use more liquid fresh water per year than is supplied each year by rain or by that portion of the snowfall that eventually melts. If all the *precipitated* liquid fresh water is divided among the 4 billion people on Earth right now, each would get 30,000 cubic meters (8,000,000 gallons) each year. That is enough to support 20 times the present population on Earth.

But wait! The liquid fresh water of Earth is *not* evenly spread among the world's population. Nor does the rain fall evenly, either in space or in time. The result is that some areas of the world have too much water while other areas of the world have too little. There are rain forests and there are deserts; there are times when there are disastrous floods and other times when there are disastrous droughts.

Furthermore, most of the fresh water on Earth makes its way back to the sea without a reasonable chance of being used by human beings at all; and much of the fresh water that we could use is being polluted—more all the time. The result is that, amazing to say in this water-logged planet of ours, we are heading rapidly into a disastrous worldwide water shortage.

Well, then, what do we do?

1. Obviously, we must, most of all, control population. If we multiply the world's population by twenty times—something we can do in 150 years if we put our minds to it—our needs will outrun the total rain supply.

2. We must do nothing to destroy the fresh water available to us. We must minimize pollution and we must avoid destroying the soil by unwise agricultural practices that lower its ability to store water, thus promoting the spread of deserts.

3. We must minimize waste and must make use of our fresh water supply more efficiently. For instance, the Amazon River, the largest in the world, discharges into the sea in one year 7200 cubic kilometers (1700 cubic miles) of fresh water, enough to supply the needs of the present population of the world indefinitely—but virtually none of it is used by man. On the other hand, we mustn't overuse the fresh water, either. We mustn't tap ground water, for instance, at a rate faster than it can be replaced, for the dropping of the ground-water level or its invasion by salt water could be ruinous.

4. Water must be viewed as a global resource and efforts must be made to transfer it from points of excess to points of deficiency, as we routinely do for food and fuel, for instance.

So much for making do with what we have. Is there any way in which we can increase the supply? Well—

a. We can minimize the loss of fresh water by evaporation, by placing single-molecule films of certain solid alcohols, or layers of small plastic balls, on exposed water surfaces. Such evaporation barriers are difficult to maintain, however, since wind and wave tend to break them up. And if they *are* maintained, they may interfere with the oxygenation of the water below.

b. Any rain that falls on the ocean is completely wasted. It might better fall on the land—it will then, in any case, return to the ocean, but it can be used en route. Any method of weather control we could devise that would shift the rain from sea to land would be helpful.

c. Since the ultimate source of rain is the evaporation of sea water by solar heat, we can add our human effort in that direction and get fresh water by desalinating ocean water artificially. This is not a blue-sky project but is done routinely today. Large ships get their fresh water by desalination, and energy-rich, water-poor nations such as Kuwait and Saudi Arabia do so, too—and are planning expansions of such equipment in the future. This *does* take energy in large amounts, however, and, at the moment, we are ill-equipped to commit those large amounts. Is there anything else?

Well, as I noted above, 98 percent of the fresh water supply on Earth is in the form of ice, which need only be melted, not distilled. Melting would take much less energy than desalination does.

The major trouble is that the ice is chiefly in Greenland and Antarctica and is not very accessible.

Some of the ice, however, is floating on the ocean. Can icebergs be dragged to where water is needed without increasing the cost to prohibitive levels?

The Arctic icebergs of the North Atlantic are relatively far away from most of those regions on Earth that are most in need of water. They would have to be moved around Africa, for instance, to reach the Middle East and around South America to reach the American West.

But what about the huge tabular icebergs of the Antarctic? These could be moved directly northward to desiccated areas without having to dodge continental land masses. And even a relatively small iceberg of this type would represent 100,000,000 cubic meters of fresh water, or a year's supply for 67,000 people.

Such an iceberg would have to be dragged slowly northward to the Middle East, say, right through the warm waters of the tropics. The iceberg would have to be trimmed to a shiplike form to reduce water resistance; it would have to be insulated on the sides and bottom to reduce melting; and once it reached Middle East waters, it would have to be sliced up, each slice melted, and the water stored.

Can all this be done without making the expense of iceberg water greater than that of desalination water? Some experts think so, and I look forward to seeing the attempt made.

After all, how better to avenge the *Titanic* than to put icebergs to such a vital use?

6. By Land and By Sea

I am writing my autobiography* and it is running longer than I had expected. This is not because my life has been exciting and full of incident, or because it has been wound up with the great events of the world, or because I have known great people and been involved in great causes. It's just that I do tend to run on and on when I am on my favorite subject.

The result is that I am consuming paper at a great rate, and the other day, when I stopped at a stationery store to buy a ream of onionskin, it occurred to me that it was silly to make the trip so often. I therefore said to the young man behind the counter, pointing upward, "Let me have two reams of onionskin—that kind you've got up there near the ceiling."

He fetched a ladder, scrambled up it, seized a ream, and started down. I cried out to him, "No, no, *two* reams, *two* reams—one, two." I held up two fingers to help him get the idea.

He hesitated. *Two* reams must have seemed to him to be against nature. But he took the second ream and brought both down slowly. As I paid for them, he looked at me curiously, as though trying to penetrate the strange motivations that led me to such an outlandish purchase, and said, "Say, what are you doing? Writing a book?"

* Since this essay was first written, the autobiography has been finished and the first volume published—*In Memory Yet Green* (Doubleday, 1979).

"I have been known to do so," I said austerely and left. I thought it was a very moderate answer considering that the number of my published books was at that moment, 188.

I *am* writing a book. For over a quarter of a century I have been writing a book on any day you care to name. Every time I write one of these essays I write one-seventeenth of a book, for every seventeen of them are collected into one by the esteemed people at Doubleday.

What's more, I'll never run out of subjects, because anything at all can inspire one. Did I discuss Antarctic icebergs in the previous chapter? Well, then, let's discuss in this one how they came to be discovered and when. And, of course, I'll begin at the beginning.

The earliest hominids seem to have evolved, perhaps as long as 10,000,000 years ago, in east-central Africa, roughly where the Equator is. From that starting point, the hominids evolved and spread outward—ever outward.

If we were to reason strictly from Earth's spherical nature, we would suppose that the hominids would spread northward and southward to an equal extent. In moving away from the Equator, it makes no difference whether you go north or south; the Earth is symmetrical in that respect. Even a rotating sphere, revolving about the Sun, with a tipped axis, is symmetrical north and south.

The accident of land formation, however, introduces an asymmetry. Africa narrows and comes to an end fairly soon as we move south of the Equator. To the north, it broadens out and is connected by a land bridge to the even vaster continent of Eurasia.

The amount of land which primitive hominids could encounter when moving north of the Equator is some seven times as great as that which they would encounter in moving south of the Equator.

Even if we allow for the fact that much of the northern lands are too cold for hominids evolved in the tropics, the potential drift has been northward rather than southward.

This accident of geography has affected humanity all

through history, so that civilizations tended to arise in the Northern Hemisphere rather than the Southern and exploration penetrated the Arctic long before the Antarctic.

What I am going to do then is to consider the southward push, the harder direction for human expansion, and see by what steps it took place even though—I warn you—I may take the long way round to do so.

The hominid ancestors of man did travel southward into South Africa, of course, and fossils of the Australopithecines were first found there in 1924.

The southernmost point of the African continent is Cape Agulhas which is 160 kilometers (100 miles) southeast of Capetown. It is quite possible that some small Australopithecine may have stood on that point of land more than a million years ago. If so, that would represent the farthest south any hominid predecessor of *Homo sapiens* ever reached. It is not very far south, to be sure, for it is only latitude 34.5° S., about as far south of the Equator as Beirut, Lebanon, or Wilmington, North Carolina, is north of it.

But then, no hominid before *Homo sapiens* was ever capable of crossing a significant stretch of open sea. This means that the early hominids were confined to the "World Island"—Africa, Asia, and Europe. In addition, they reached the westernmost islands of Indonesia by getting across the narrow straits separating the Malay Peninsula from Sumatra, and Sumatra from Java.

The World Island was undoubtedly very thinly populated by these early hominids, but *Homo erectus,* who flourished between 500,000 and 1,000,000 years ago, existed in northern China and in Indonesia, where their fossil remains received such names as Peking Man and Java Man.

Homo sapiens (which includes Neanderthal Man as well as ourselves) was in existence at least as long as 350,000 years ago and eventually became the only hominid existing, either through default or by actually hunting down his more primitive cousins.

When it came time for hominids to spread beyond the

confines of the World Island, it was *Homo sapiens* that did so.

Prehistoric *Homo sapiens* was no better at crossing stretches of sea than his hominid predecessors were, but he managed to take advantage of land bridges that formed when the sea level dropped in glacial periods. Perhaps 25,000 years ago, human beings managed to make their way across the islands of the Indonesian archipelago and finally reached New Guinea and Australia.

The invaders worked their way southward, and when they reached the southeastern corner of Australia (what is now the state of Victoria), they broke the hominid record for southern penetration, for that part of the continent extends farther south than southernmost Africa. Indeed, Southeast Cape, the southernmost point of Australia, is at latitude 39.08° S., about as far south of the Equator as Valencia, Spain, or Washington, D.C., is north.

That does not represent the true record in that area of the world, though. The island of Tasmania is 200 kilometers (125 miles) southeast of Australia and this, too, was reached and occupied by the invaders. The southernmost part of Tasmania is also called Southeast Cape, and that is at latitude 43.48° S., as far south of the Equator as Florence, Italy, or Toronto, Canada, is north.

At the time that Australia was reached by immigrants from southeastern Asia, North America was reached by immigrants from northeastern Asia.

The human beings entering North America worked their way southward the full 16,500-kilometer (10,250-mile) length of the two American continents and may, by 8000 B.C., have reached the southern tip of South America.

As it happens, the southernmost 1100 kilometers (700 miles) of the narrowing horn of South America (a region called Patagonia) is farther south than any other inhabited continental area of the world.

The southernmost point of South America is nearly at latitude 54° S., which is as far south of the Equator as Hamburg, Germany or Edmonton, Alberta, is north.

South of this southern tip of South America is a large island surrounded by a haze of smaller islands which we can lump together as Tierra del Fuego. Its southernmost point is Cabo de Hornos (Cape Horn) which is at latitude 56.00° S., as far south of the Equator as Edinburgh, Scotland, or the southernmost bit of the Alaska panhandle, is north.

Tierra del Fuego is the southernmost bit of land that, even to this day, is permanently occupied by human beings.

To summarize, then, whereas the predecessors of *Homo sapiens* were restricted to the World Island and reached no farther south than the southern tip of Africa, *Homo sapiens*, in prehistoric times, penetrated both Australia and the Americas, reaching farther south than Africa in both new continents and attaining Tierra del Fuego at about the time that cities were beginning to arise in the Near East and what we call "civilization" was approaching birth.

These penetrations of *Homo sapiens,* up to 8000 B.C., however, were essentially the kind of expansions that roving tribes could achieve on foot. They spread out over continuous land or over streams that could be forded or over arms of the sea that could be crossed in glacial times. Any bits of land that remained separated from the continental areas by sizable widths of ocean remained human-free even as late as Roman times.

The final exploration of the Earth had to depend on the development of ships capable of extended voyages, on the coming of maritime development. After 8000 B.C., human expansion was by sea rather than by land.

The first maritime people were the Minoan Cretans, who developed a navy as early as 3000 B.C., but their activities were confined to the eastern Mediterranean.

The next group of remarkable mariners were the Phoenicians who reached their peak after 1000 B.C., when they burst out of the Mediterranean and became the first to maneuver their ships long distances across open ocean. They reached the British Isles to the north and did better than that to the south.

The Greek historian Herodotus, writing about 430 B.C., tells us that about 600 B.C. the Egyptian Pharaoh, Necho, sent out a fleet of ships, manned by Phoenicians, to explore the African coast. The ships apparently sailed down the Red Sea and returned by way of the Strait of Gibraltar in a voyage that lasted three years.

Herodotus haughtily refuses to believe one story told by the Phoenicians. He says, "These men made a statement which I do not myself believe, though others may, to the effect that as they sailed on a westerly course round the southern end of Africa, they had the sun on their right—to northward of them."

Herodotus knew that the noonday Sun was always to the south of the zenith in Greece and in all the lands that he had visited, and he considered this a law of nature, apparently. Yet if the Phoenicians had really rounded Africa, they would have penetrated into the South Temperate Zone and the noonday Sun would indeed have been to the north of zenith. It is precisely the story Herodotus would not believe that makes us confident that the Phoenicians were telling the truth—for they could not have dreamed of a northward Sun (at a time when Earth was not yet perceived to be a globe) without having actually seen it.

Nothing came of this voyage, however, and the next group of great mariners in the Atlantic were the Vikings, fourteen centuries later. After A.D. 800, the Viking sea raiders terrorized the coasts of Europe and carried their explorations westward and northwestward into Arctic regions. They settled Iceland permanently and Greenland temporarily and seem to have reached the coast of North America. (Meanwhile, the Eskimos, in more traditional landbound fashion, moving east from Asia, were colonizing the northern shores of North America and Greenland.)

The Vikings, however, remained in the Northern Hemisphere. It is the southward exploration—the harder direction—that concerns me here.

On the shores of the Pacific and Indian oceans there were Chinese and Arab vessels which penetrated the oceans for purposes of trade, but their deeds were dwarfed by those of the Pacific Island people who, considering the technology at their disposal, were far and away the most daring and accomplished seafarers the world has ever seen.

Starting from New Guinea, human beings must have managed to reach some of the smaller islands to the east—the Solomon Islands, New Caledonia, and so on, which make up the region now known as Melanesia.

Then from Melanesia, about A.D. 300, certain islanders broke out into the vast Pacific and, over the next thousand years, crisscrossed the ocean and settled virtually every island in it. In the end, they had occupied an enormous Pacific triangle covering 14,000,000 square kilometers (5,500,000 square miles), a triangle that was almost all water, of course. This area is now called Polynesia ("many islands"—from the Greek) and the navigators are Polynesians.

The drift was mainly northward into the warmer waters. The northern apex of the triangle was Hawaii and the eastern apex Easter Island.

It was not till early medieval times (in Europe) that the most significant penetration was made by the Polynesians. Until then, the large islands of New Zealand had remained human-free; the largest pieces of temperate land remaining empty that late in history. (In fact, except for bats, it was mammal-free, which made it a paradise for wingless birds such as the moa.)

In A.D. 800, however, a Polynesian people reached New Zealand and established the southern apex of the Polynesian triangle (and within three centuries had wiped out the moas).

New Zealand consists of two main islands, North Island and South Island. To the south of South Island is Stewart Island, whose southernmost point is at latitude 47.17° S., or as far south of the Equator as Nantes, France, or Tacoma, Washington, is north.

Until the dawn of modern times, the colonization of New Zealand by the Polynesians represents the southernmost penetration of maritime humanity. New Zealand reaches farther south than any point in Africa or even

Tasmania, but falls well short of the southernmost portion of South America.

Nevertheless, by the end of the Middle Ages in Europe, all the significant land areas of the world were populated except for those which, by virtue of their sheer physical hostility to human beings (deserts, mountain tops, polar regions), could not be inhabited.

Europe and the Orient had been trading even in ancient times by means of slow land caravans and through many middlemen, with the result that silk, pepper, and other Oriental commodities reached Europe. The European nations on the Atlantic shores were at the end of the line and the Oriental commodities they wanted were increasingly hard to get, chiefly because of the activities of the middlemen, some of whom, like the Turks, were actively hostile and some of whom, like the Venetians, were merely greedy.

The result was that some Westerners conceived the notion of skipping all the middlemen and dealing with the Orient directly. The only way this could be done was by sea and that meant going around the World Island.

This could be done by either skirting its northern shores (the "northeast passage") or its southern shores (the "southeast passage"), and the trouble was that no West European had ever done either. It wasn't known whether those shores were practical lanes for seafaring men or, in fact, whether they existed at all. It might well be that there was no sea connection at all between western Europe and eastern Asia. After all, there might be two landlocked oceans on Earth, one off the shores of western Europe and one off the shores of eastern Asia.

(The Phoenician voyage of 600 B.C. was a strong indication that there was a single ocean, but only those who read Herodotus knew of it and Herodotus said he didn't believe it.)

The hunger for Oriental commodities drove West Europeans to ocean exploration, nevertheless, to see what the situation might be. Here they were mightily helped by the use of the mariner's compass, something the Phoenicians,

Vikings, and Polynesians did not have and which came into use in Europe in the 1200s.

The compass, by the way, was first developed by the Chinese, who, in the European Middle Ages, led the world in marine technology. The Chinese, however, oversatisfied with their own highly refined and subtle civilization, deliberately chose isolation and put an end to their ocean voyages. They let the world pass as unworthy of their notice.

The result was that eventually the world intruded on them, in the form of Western ships armed with the compass and with gunpowder, which Europeans had taken from the Chinese and improved. The Chinese had to pay for their snub of technology with some two centuries of humiliation. —There's a moral there, but I won't bother pointing it out.

Of the European nations, Portugal led the way. The northeast passage, assuming it existed, didn't look promising. It would mean passing around the Scandinavian coast, which was known to be inhospitable and there was the expectation of worse beyond. The southeast passage, however, was completely unknown and might, therefore, be infinitely the superior.† For the first time, apparently, south was the easier direction.

Portugal's guiding genius in this respect was King John I's younger son, Prince Henry (known in history as "Henry the Navigator").

The Portuguese were fighting the Muslims on African soil, and in 1415 Henry took part in a battle at Ceuta, on the northwestern tip of Africa, and was knighted for heroism. Although he himself never penetrated deeper into Africa, he fell in love with the continent, so to speak, and with the project of exploring its coasts. He dedicated his life to working out the southeast passage.

He established an observatory and school for navigation

---

† And possibly not. Some people felt that the equatorial zones were inpenetrable zones of insupportable heat.

at Sagres on Cape St. Vincent in 1418. This was in southernmost Portugal at the southwestern tip of Europe. Year after year he outfitted and sent out ships that inched their way farther and farther down the West African coast. He even supervised the collection of astronomical data to ensure the greater safety and success of the ships.

The effort was, allowing for the technological level of the times, rather the equivalent of the effort to put a man on the Moon in our own times. Things moved more slowly then than now, however, and by the time Henry the Navigator died, in 1460, his ships had only reached the westernmost bulge of Africa, where Dakar is now situated, and were only about one fifth of the way around the continent.

The effort survived Henry's death, however. Portuguese ships kept probing farther south, looking for a place where the coast would turn eastward. For a while they thought they had it when, at the southern end of what is now Liberia, the coast turned eastward for 1500 kilometers (900 miles) and then turned—southward again.

Finally, in August 1487, the Portuguese navigator Bartolomeu Dias set sail with three ships. Having gone farther south than any Portuguese navigator before him, he was caught in a storm and driven farther south still. When the storm let up, he found himself out of sight of land. He sailed northward and on February 3, 1488, reached a coast that was going east and west. He sailed east long enough to see that it was going to turn northward again.

Satisfied that he had rounded the southernmost part of Africa, he backtracked and named the point of land which marked the eastward turn of the coast the "Cape of Storms." King John II of Portugal, on hearing the news when Dias returned, more properly called it the "Cape of Good Hope." (It is not quite the most southerly point of the continent.)

This marked the beginning of the West European domination of the sea-lanes, a domination that continued for four and a half centuries. Achieving the southeast passage was now just a matter of perseverance (but Dias had not yet equaled the southernmost penetration of the Polynesian mariners in New Zealand).

It was clear that Africa extended so far southward that the southeast passage would represent a sailing length of about 20,000 kilometers (12,500 miles). It seemed to some, notably to Christopher Columbus of Genoa, that, since the world was a globe, it would be simpler to reach the eastern coast of Asia by sailing due west. Columbus believed that the Earth was 18,000 miles in circumference and that eastern Asia extended farther east than it did (following errors in Ptolemy's book on astronomy which was then fifteen centuries old). Columbus felt, therefore, that a westward journey of some 5000 kilometers (3000 miles) would reach the Asian coast and that this was only a quarter the length of the southeast passage.

He couldn't sell his idea to the Portuguese, however. They suspected the Earth was 25,000 miles in circumference, basing this on their greater navigational experience. They therefore felt that the "western passage" would be nearly as long as the southeast passage. Furthermore, the western passage would be out of sight of land all the way, while the southeast passage would follow coasts and find havens most of the way.

The Spanish monarchs, however, after being badgered unmercifully by Columbus, decided to get rid of him. They sent him on his way in 1492 with three derelict ships and a convict crew, fully expecting never to see him again.

But we know what happened. Columbus returned in triumph and Spain was on its way to a period in which it would be the strongest nation on Earth. The Portuguese king was sporting enough to congratulate Columbus, but he could afford that. He was probably certain that the new shores, whatever they were, were not Asia and that he would still beat Spain to the punch as far as trading with the Orient was concerned—and he did.

In November 1497 the Portuguese navigator Vasco da Gama rounded Africa and made his way up its eastern coast, striking eastward across the Arabian Sea and landing in Calicut on the southwestern coast of India in May 1498. The southeast passage had been achieved.

It was on Da Gama's voyage, by the way, that the length of the trip and the nature of the sailors' diet finally managed to deplete them of vitamin C and bring on the first known attack of shipboard scurvy, something that was to plague long-distance mariners for two and a half centuries.

The Portuguese were able, as a result of their explorations, to build up the first European overseas empires and to take non-European slaves, a practice other European nations were to follow for centuries, till all gave up those empires—the Portuguese last.

Columbus died in 1506, insisting to the end that he had reached Asia, but the suspicion grew steadily that he had not.

In 1504 an Italian navigator, with the Latinized name of Americus Vespucius, first published arguments to the effect that the land Columbus had discovered was a hitherto unknown continent and that Asia lay still farther westward behind a *second* ocean.

In 1507 a German geographer, Martin Waldseemüller, accepted this notion, and marked off the new continent and the second ocean on a map he drew. He suggested the new continent be named America in honor of Vespucius and this, in my opinion, was a just suggestion.

A Spanish explorer, Vasco Núñez de Balboa, was living on the Atlantic shore of what we now call Panama, a place that was one of the centers of early Spanish colonization of the Americas. Balboa had no notion that he was living on a narrow isthmus, of course.

On September 1, 1513, he organized an expedition inland to search for gold, for he was badly in debt. On September 25, after topping a rise, he found himself gazing at a wide expanse of water with no sign of an opposite shore. It could not be the Atlantic Ocean because the Atlantic coastline nowhere curved so as to get in front of him like that. Balboa called what he saw the "South Sea" because at that point it lay to the south of the shoreline.

Balboa could not tell in that one glance that he had discovered the second ocean concerning which Vespucius had theorized, but he had. Further explorations made it certain, and now it seemed that to reach Asia from Europe by sea going westward, one had to skirt the Americas either on the north or on the south and then keep on going. In other words, there was now the possibility of a "northwest passage" and a "southwest passage."

You might ask, Why bother?—the southeast passage was practical and effective. . . .

Ah, but the Portuguese had a stranglehold on that and were not disposed to share it. If the Spaniards wanted to cut in on the lucrative trade with the Orient, they had to move through either the northwest passage or the southwest passage, and again there was the question of whether either was navigable or whether either existed at all.

How the Spaniards managed will be discussed in the next chapter.

7. We Were the First That Ever Burst —

It is well known that I do not take airplanes and that I hate to travel, and yet such is the pervasive mobility of human beings in our society that I have, on occasion, been thousands of kilometers from home. Since I'm writing a series of articles that deals essentially with the exploration of the Earth, I feel impelled to list the bounds of my personal wanderings thereupon.

The farthest east I have ever been is Moscow (longitude 37° E.) and that happens to be the farthest north I have ever been, too (latitude 55° N.). That, however, was at the age of not quite three, when my parents were in the process of getting their tickets to come to the United States and were planning to take me with them.

That, of course, was an involuntary trip, and I remember nothing of it.

Where only voluntary trips are concerned, then the farthest east *and* north I have ever been is London (0.1° W. and 51° N.), in 1974.

If we consider my London visit as my farthest penetration northeast, then the farthest southeast I have ever been came in 1973 when I was on the ship *Canberra*, anchored off the city of Dakar in Senegal (17°W. and 14° N.).

Directly southward, my furthest reach came in 1975, when I actually set foot on South America, at La Guaira, Venezuela (10° N.).

As for westward, my record penetration came in 1946 when I reached Honolulu, Hawaii (157° W.). That trip to Honolulu, however, came when I was in the Army and, again, that was not voluntary. My farthest trip westward under my own steam was to Chicago (87° W.), which I reached for the first time in 1952.*

As you see, despite all these daring trips of mine (from all of which I longed to return home), I have never crossed the Equator, nor, since I came to the United States, have I ever crossed the Prime Meridian. I have never seen any part of Asia or Australia and, in particular, I have never been within 10,000 kilometers of Antarctica, which is the central subject of this article.

In the previous chapter, while discussing the southward penetration of humanity, I had reached the point where Columbus had successfully crossed the Atlantic Ocean and reached land and where it had come to be realized that a second ocean separated that new land from eastern Asia (which had been Columbus's goal).

In 1493, in an agreement mediated by the Pope, Spain and Portugal had divided up the non-European portions of the Earth between themselves. Spain was to have everything west of a line running down the mid-Atlantic, Portuguese everything east. Neither side thought to have the line run all the way around the globe. It ran from the North Pole to the South Pole on the Atlantic side only.

It turned out that on the Portuguese side of the line was the coveted Far East, while on the Spanish side of the line were only the primitive American continents. Spain chafed a lot over this.

In 1517 a Portuguese navigator, Ferdinand Magellan, having been basely mistreated (as he thought) by his government, defected to Spain. He suggested to the Spanish King that it was perfectly possible to stay west of the Line of Demarcation and still reach the Far East—if only one went far enough. All one had to do was to move around the Americas, either north or south, and then keep going.

---

* In 1978, after this was written, I reached San Francisco, which is 122° W.

There was no Line of Demarcation on the other side to stop them.

On September 20, 1519, Magellan left Spain with five ships. He might have tried skirting the Americas to the north, but it was already known that if there were sea-lanes on its northern boundary (the northwest passage), they were in Arctic waters that would be difficult to navigate. As to where the Americas might end in the south, no one as yet had any information, and the southwest passage might prove easy.

Magellan gambled on that. He led his ships down the eastern coast of South America, probing into every hopeful inlet, such as the ones where Rio de Janeiro and Buenos Aires are now located.

Farther south, Magellan and his men came across Indian natives who seemed, to them, to have big feet, and to this day the narrowing southernmost reach of South America is called Patagonia, which is "big feet" in Spanish.

Finally, on October 21, 1520, when Magellan had gone far enough south to find the weather semi-Arctic (or, rather, semi-Antarctic) in nature, he found an inlet which seemed promising. He made his way through it, wondering if it might not be a dead end—and the passage is called the Strait of Magellan to this day.

When Magellan came out into the open ocean at last and found it sunny and calm (at least at that moment), he looked out over the vast, flat expanse and, with tears running down his cheek, named it the Pacific (peaceful) Ocean. Of the European mariners, he was the first that ever burst into the Pacific Ocean.†

Magellan went on to push across the Pacific Ocean in a harrowing voyage during which he went ninety-nine days

† From their far southern position, the mariners observed two luminous patches in the heavens that looked like detached portions of the Milky Way and that could not be seen from Europe. They have been termed the "Magellanic Clouds" ever since and they are two satellite galaxies of the Milky Way.

without sight of land. Magellan himself was killed by natives in the Philippine Islands, but he reached the Far East as he said he would. The ships went on, picking up cargo and making their way through the Indian Ocean and around Africa, until, after a three-year voyage, one ship only, with no more than eighteen men on board under the command of Juan Sebastián del Cano, made it back to Spain. They were the first men to have circumnavigated the globe, and the spices they brought back far more than paid the expense of the voyage, if one does not count the loss of life.

From the standpoint of this chapter, however, what was most interesting about the voyage was that when Magellan and his men went through the Strait of Magellan, they penetrated farther south than any Europeans in history, reaching 54° S.

They did not, however, break the record in an absolute sense. South of the strait was land. The mariners sighted campfires on its shores and called it Tierra del Fuego (Land of Fire), a name it bears to this day. There were Indians on Tierra del Fuego, who had penetrated farther south than Magellan's reach, thousands of years before Magellan.

No one cared about Tierra del Fuego. There was clearly nothing to be gained from that frigid and dank region. It was just a place one passed by in taking the southwest passage to the glorious East. It was somehow assumed, on the basis of no evidence whatever, that it was the tip of another continent. Early maps showed this, with much imaginary coastal detail.

About fifty years after Magellan's voyage, the Spaniards were looting what is now called Latin America rather ruthlessly, and the English, who were in a cold war with Spain, were taking Spanish ships and raiding coastal settlements whenever they could, in order to loot the looters. It was all unofficial, for Queen Elizabeth I professed herself horrified at the action of the English raiders (which frustrated the Spaniards no end, since she shared in the spoils and knighted the spoilers).

The most successful of the English looters was Francis Drake. It occurred to Drake that the Spaniards were well fortified on the Atlantic side of their American possessions, but utterly asleep on the Pacific side where they lived in fancied security.

Toward the end of 1577, therefore, Drake decided to take his ships through the Strait of Magellan and up the western shores of America. This he did, reaching as far north as what is now San Francisco Bay, and gathering so much loot that he stopped only because additional material would founder his vessel, the *Golden Hind*.

He then headed across the Pacific and became the second mariner to circumnavigate the globe.

From the standpoint of this chapter, however, what was interesting about this voyage was that after Drake emerged from the Strait of Magellan, he found the Pacific to be not at all pacific. In fact, he was struck by a violent storm that drove him southward quite against his will. It drove him sufficiently far south for him to be able to demonstrate that Tierra del Fuego was an island and that south of it there lay open ocean. That stretch of open ocean has been called Drake Passage ever since.

Drake, in being driven as far south as latitude 58° S., set a human all-time record for southward penetration. No one in all the history of the human species had ever been so far south.

If Tierra del Fuego hadn't seemed worth investigating, the possibilities of still less inviting land south of Drake Passage certainly aroused no curiosity. The thrust of exploration lay in other directions—say, in the location of additional land in the neighborhood of the Indonesian islands, which were fabulously wealthy in the raw materials that Europeans desired.

In 1606 a Spanish navigator, Luis Vaez de Torres, sailed south of the largest of the Indonesian islands, New Guinea, and the water passage there is now called Torres Strait.

In 1602, the Dutch had begun to gain control of the

Indonesian islands, ousting the Portuguese. As vague reports came in of land south of Torres Strait, Anton Van Diemen, the governor general of the islands, took action. In 1642 he sent an exploring expedition southward under the Dutch explorer Abel Janszoon Tasman.

Tasman had amazingly bad fortune. In the course of a ten-month voyage, he managed to sail all around an island almost as large as the United States without sighting it or suspecting its existence. He did come across a small island to the southeast of the larger one and called it Van Diemen's Land after his boss. It is now, more justly, known as Tasmania, after the explorer. Tasman also located a pair of larger islands to the southeast of Tasmania, which he named after the Dutch province, Zeeland, and which are known as New Zealand now.

In a later voyage, Tasman spotted the northern shores of the large island he had missed and called it New Holland. He had no suspicion that it was continental in size.

None of Tasman's discoveries were as far south as the southern tip of South America, so he set no records for southern penetration.

There remained a glittering prize to be found, though, and that was an imaginary land invented by the Greeks. Some Greeks, who knew the Earth was round and that all the land they knew was in the Northern Hemisphere, felt that by the principle of symmetry there ought to be an equal amount of land in the Southern Hemisphere. This land came to be called Terra Australis (Southern Land).

In 1700 the known land south of the Equator included the southern two thirds of South America, the southern third of Africa, and assorted islands, of which the largest was New Guinea. All of this put together was considerably smaller in area than the northern land, so it seemed there must still be something missing.

Suspicion centered on the Pacific Ocean. The known shores of the Americas, of Asia, and of the Indonesian islands enclosed an area that made up half the surface of the Earth. Surely, it couldn't be empty of land.

European explorers crisscrossed the vast Pacific looking for Terra Australis and some of them probed southward.

In 1738 a French naval officer, Pierre Bouvet de Lozier, sailed along the latitude of 55° S. for 1500 kilometers (930 miles), but all he found was a small island about 2600 kilometers (1600 miles) south of the southern tip of Africa. This island, now called "Bouvet Island," is not quite as far as the southern tip of South America so it did not set a record for southerly penetration.

Another French navigator, Yves Joseph de Kerguelen-Trémarec, set out in 1771 and found another island, larger than Bouvet Island, about 4000 kilometers (2500 miles) southeast of the southern tip of Africa and not quite as far south as Bouvet Island. The new island is now called "Kerguelen Island."

The greatest of all the Pacific explorers, however, was James Cook, who made so remarkable a name for himself as a ship's master that he is universally known as "Captain Cook" and his first name is almost forgotten.

Between 1768 and 1771, on the first of three great voyages, Captain Cook sailed across the South Pacific Ocean and explored the coasts of New Guinea and New Zealand. He also investigated the bounds of Tasman's New Holland and showed it to be an island—but an island almost as large as all of Europe. He called it "Australia," a clear reminiscence of the legendary Terra Australis.

Even with Australia subtracted, the Pacific Ocean seemed impossibly large, and in 1772 Captain Cook set off on a second expedition. He scoured the Pacific Ocean so thoroughly that it became quite plain that nothing of continental size could exist between Australia and South America. In fact, said Cook rather mournfully, if any southern continent existed, it would have to be so far south that it would be worthless.

Of course, the Pacific Ocean isn't entirely empty. The land it contains is nearly 1,000,000 square kilometers (400,000 square miles) in area, or half again as large as Texas. The trouble is that it is split into some 10,000 tiny islands.

In his third and last voyage, from 1776 to 1779‡ he explored the northern Pacific and was finally killed (and eaten) by the inhabitants of the Hawaiian Islands.

In the course of his second voyage, however, Cook finally surpassed Drake's two hundred-year-old record southerly penetration.

On January 17, 1773, Captain Cook's ship reached a latitude of 66.5° S. and crossed the Antarctic Circle. It was the first time *any* human being had crossed the Antarctic Circle. He made two other crossings in the course of his journey and his most southerly penetration took place on January 30, 1774, when he reached latitude 71.17° S.* He was then only 1820 kilometers (1130 miles) from the South Pole.

Captain Cook saw no continental land on any of his southern penetrations. His ship was always stopped by massed ice, and for all he could tell, it was ice all the way to the South Pole, with no land at all.

The Antarctic penetrations mentioned in this article must have been on the mind of Samuel Taylor Coleridge in 1798, when he wrote *The Rime of the Ancient Mariner*. The Ancient Mariner leaves England, and his ship is driven southward through the South Atlantic (as the storm had driven Drake) until, like Cook, he reached the southern ice:

> *The ice was here, the ice was there,*
> *The ice was all around:*
> *It cracked and growled, and roared and howled,*
> *Like noises in a swound!*

Under the guidance of an albatross, the explorers find their way out of the ice northward again and into the Pa-

---

‡ During this voyage, the American privateers left him severely alone despite the fact that we were in rebellion against Great Britain. Cook's voyages were recognized as important enough to transcend national quarrels.

* Please remember that January and February are the height of the Antarctic summer.

cific Ocean on the other side of South America—and here there is a reminiscence of Magellan's great penetration:

> *The fair breeze blew, the white foam flew,*
> *The furrow followed free;*
> *We were the first that ever burst*
> *Into that silent sea.*

Captain Cook did discover some southern islands, though. He discovered South Georgia Island (named for George III), which is just about as far south as Tierra del Fuego, but 1750 kilometers (1100 miles) to the east. He also discovered the South Sandwich Islands, which he named for John Montagu, fourth Earl of Sandwich and First Lord of the Admiralty.†

The South Sandwich Islands lie to the southeast of South Georgia. They are in a north–south line and represent the first pieces of land ever discovered lying south of Tierra del Fuego. The southernmost of the South Sandwich Islands, appropriately called "Cook Island," is at latitude 59.3° S.

One consequence of Captain Cook's explorations was that the Antarctic waters were shown to be rich in seals and whales. Ships known as "sealers" and "whalers" went south to prey on those mammals, and new discoveries were made in consequence.

In October 1819 a British navigator, William Smith, discovered the South Shetland Islands. These are directly south of Tierra del Fuego and set a new southerly record for land. The southernmost of the group, appropriately called Smith Island, is at latitude 63.0° S.

Then the British monopoly on Antarctic exploration was broken. In 1819 a Russian explorer, Fabian Gottlieb Bellingshausen, was sent southward by Tsar Alexander I, with

---

† The "sandwich" we eat is also named for him, as are the Sandwich Islands, as Cook called what we now call the Hawaiian Islands.

specific instructions to better Cook's record southern penetration.

Bellingshausen didn't, but he did come across a small island, about the size and shape of Manhattan, which he named Peter I Island after Tsar Peter the Great. That island is at latitude 68.8° S., or 240 kilometers (150 miles) *south* of the Antarctic Circle. It was the first piece of truly Antarctic land ever discovered.

Bellingshausen went on to discover a large island about 600 kilometers (370 miles) further west and this he named for Tsar Alexander. Bellingshausen thought it might be part of a continental mass but it wasn't; it's just so wedged in with ice that its island nature is hard to demonstrate. The most southerly portion of Alexander Island is at latitude 72.5° S.

The portion of the ocean between Peter I Island and Alexander Island is known as Bellingshausen Sea.

Meanwhile, a British naval commander, Edward Bransfield, charted the South Shetland Islands and explored the waters to the south, a region still called "Bransfield Strait" in his honor. Bransfield also reported on January 30, 1820, a rather dubious sighting of land to the south of the strait, since called Graham Land, after a First Lord of the Admiralty, Sir James R. G. Graham.

On November 16, 1820, a twenty-one-year-old American sealer, Nathaniel Brown Palmer, in command of a small sloop that was part of a larger fleet, *definitely* sighted land south of Bransfield Strait.

The result was a stubborn long-lived geographical dispute between Great Britain and the United States. The British called the new land Graham Land and the Americans called it Palmer Land. There was no reason to think that the new land, whatever its name, was anything but another island, but in the end it turned out to be part of the Antarctic continental land mass, the most northerly part, a gently curving S-shaped piece of land about 1500 kilometers long, with its extreme point at latitude 63° S., or 480 kilometers (300 miles) north of the Antarctic Circle in the South Temperate Zone.

It was not till 1964 that the naming conflict was settled. It was then decided to call the northern part of the narrow finger of land Graham Land, the southern part Palmer Land, and to call the whole by the neutral name of "Antarctic Peninsula."

On February 7, 1821, an American sealer, John Davis, actually disembarked on the Antarctic Peninsula and therefore became the first human being to stand on Antarctica. His feat, however, went completely unknown, for it was not till 1955 that the log of his ship was discovered and the priority of his feat understood. (Davis, perhaps influenced by Bellingshausen, stated in his log his opinion that he was standing on an Antarctic continent, but he had no evidence for it—only intuition.)

Cook's mark was finally broken on February 20, 1823, when an English whaler, James Weddell, reached a mark of latitude 72.25° S. in an oceanic inlet now called "Weddell Sea." At the point where wind and ice turned him back he was within 1800 kilometers (1100 miles) of the South Pole.

Weddell suspected that the sea penetrated all the way to the South Pole and that only ice, not land, blocked the passage. If that were so, it would be likely that there was no Antarctic continent, but only, at best, a collection of Antarctic islands.

Weddell Sea lies just to the east of Antarctic Peninsula. In January 1841 a Scottish explorer, James Clark Ross, entered another oceanic inlet, some 2000 kilometers (1200 miles) west of Weddell Sea. The new inlet is now called Ross Sea.

Ross sailed south till he found himself stopped by a towering wall of ice, 60 to 100 meters (200 to 320 feet) high (that is, as high as a building of 20 to 30 stories). We know this, now, to be an enormous shelf of ice pushed off the land beyond onto the sea. It is called the "Ross Ice Shelf" and it covers an area about the size of France.

Ross sailed along the limits of the ice shelf that year and again the following year and eventually achieved a new southern mark of latitude 78.15° S., which is 1150 kilometers (720 miles) from the South Pole.

One of Ross's more interesting discoveries in his exploration of Ross Sea was Mount Erebus, an active volcano 3.7 kilometers (12,450 feet) high. It is on an island (Ross Island) and is at latitude 77.4° S., the southernmost active volcano in the world. The description of its red hot lava surrounded by eternal ice may have given Edgar Allan Poe the idea for an elaborate simile in his poem *Ulalume* written in 1847:

> *These were days when my heart was volcanic*
> *As the scoriac rivers that roll—*
> *As the lavas that restlessly roll*
> *Their sulphurous currents down Yaanek*
> *In the ultimate climes of the pole—*
> *That groan as they roll down Mount Yaanek*
> *In the realms of the Boreal Pole.*‡

The Weddell Sea is also covered by an ice shelf in its southern reaches, one that is called the Filchner Ice Shelf, after the twentieth-century German explorer Wilhelm Filchner.

If the two ice shelfs were imagined as disappearing, the southernmost shore of the Weddell Sea would be at about latitude 82° S. (900 kilometers [560 miles] from the South Pole) and the southernmost shore of the Ross Sea would be at about 86° S. (500 kilometers [310 miles] from the South Pole).

As long as the ice shelfs are there, however, an ultimate limit is set to southerly penetration by ship. Ross's mark is the best that can be done by any ship. To do better, explorers would have to take to their legs.

‡ Poe cheats, of course, by inventing the name "Yaanek" in order to get a rhyme for "volcanic." What's more, he transfers the volcano to the North Pole so he can use the adjective "Boreal." The corresponding adjective for the South Pole would be "austral" and that would stick Poe with a missing syllable.

The Antarctic discoveries I have described so far were in that part of the region south of South America and the Pacific, where the sea bit deep. Not so on the other side of the Antarctic south of Australia, Africa, and the Indian Ocean that lies between.

In 1831 the first sighting of Antarctic land south of Africa was made by an English navigator, John Biscoe. He called it "Enderby Land" after the owners of his vessel. He saw the land only from a distance, though. Ice prevented him from actually reaching it.

In 1840 a French explorer, Jules Dumont d'Urville, spied a shoreline near the Antarctic Circle south of Australia and called it Adélie Land, after his wife.

Between 1838 and 1842, the American naval officer Charles Wilkes headed an exploring expedition that eventually took him to the Antarctic. There he followed a long stretch of coastline between Enderby Land and Adélie Land, a coastline that followed the curve of the Antarctic Circle with surprising exactness. This stretch of coast, lying south of the Indian Ocean, is now known as "Wilkes Land."

Wilkes, on returning to the United States, was the first to proclaim that all the isolated discoveries of the previous quarter century could be fitted together to indicate the existence of a South Polar landmass of continental size. Wilkes then was the effective discoverer of Antarctica as a continent.

As it happens, I think that all national and ethnic chauvinisms are unlovely, but every once in a while, I can't resist—

Although the United States was still in its raw youth in the first half of the nineteenth century, it was an American, Palmer, who made the first definite sighting of any part of Antarctica. It was an American, Davis, who first set foot on any part of Antarctica. It was an American, Wilkes, who first established Antarctica as a continent and was its effective discoverer.

We were "the first that ever burst" into that frozen land.

8. Second to the Skua

The other day I visited the Smithsonian National Air and Space Museum in Washington and was fascinated by the wealth and variety of its exhibits.

It was, in a sense, a paean to space flight, and I kept wondering what the reaction would have been to an article that described the museum exhibits accurately but had been published in 1938 as a prediction of the next forty years.

It would have to describe close-up photographs of Mercury, spacesuits, control panels of spaceships, an actual spaceship in which astronauts had lived for three months, and many other things.

It would surely have seemed like a mad pipe dream and if someone had taken the role of good old hardheaded* Senator William Proxmire of Wisconsin, he would have waxed merrily sarcastic at the expense of "crazy science fiction."

Yet the most astonishing thing I saw at the museum was a celebration not of space flight but of its two centuries of prologue. It was a movie entitled *To Fly,* shown on a 50-foot-high screen. For thirty minutes I saw views of the Earth as seen from a balloon, from a man-carrying kite, from a stunting biplane, and so on. The views were an incredible vision of beauty that I know I'll never see in real life since I am determined to let nothing get me off the ground.

* I have often thought that the hardest head is one that's bone from ear to ear.

Did the vision fill me with delight and make me vow to fly? Not a bit! Incredibly beautiful though it all was, I sat white-knuckled in my seat with my intestines spontaneously forming square knots and half hitches. You can't argue with an acrophobe.

And similarly, not all the photographic views of the frozen grandeur of Antarctica ever elicits within me even the slightest desire to visit the place in person. I am content to write about it without seeing it.

In the previous chapter, I brought the history of Antarctica up to the 1840s, when explorers had nosed about its shore sufficiently to reveal the fact that what occupied the South Frigid Zone was frozen land of continental size.

While this was a brand-new extension of the human range—for until the nineteenth century no human being, no hominid even, had ever been within sight of Antarctica —it was not a new extension of life in general.

The shores of Antarctica and the ocean that washes those shores teem with life.

This is not surprising. Whereas most solids become more water-soluble as temperature rises, gases become *less* soluble as temperature rises. The freezing water of the polar regions carries 60 percent more dissolved oxygen than does the tepid water of the tropical ocean surface.

Since, except for some bacteria, life depends on oxygen, the polar ocean blooms with microscopic life, which in turn supports the larger life that feeds on it, which in turn supports still larger life, and so on.

It is in the polar regions and, particularly, in the Antarctic that you would expect to find a food supply rich enough to support the larger whales. In the Antarctic you will find the blue whale which can attain a mass of 130,000 kilograms (150 tons) and which is the largest animal that has ever existed.

Of course, whales are thoroughly aquatic and are creatures of the sea only. They are never actually inhabitants of even the edges of Antarctica.

Aquatic animals, less thoroughly adapted to the sea,

must come out on land at least to breed, and for some of these the Antarctica shores are the breeding ground.

Of the forty-seven species of seals, for instance, five are native to the Antarctica shores. The most common is the crabeater seal (which does not eat crabs).

More spectacular, though, is the leopard seal which is the most dangerous carnivore of the family. The leopard seals will eat sizable birds and fish without discrimination and need fear no enemy of their own other than the still larger, and more dangerous, killer whale (which is, in turn, altogether immune to predation, barring the interference of human beings).

The most Antarctic of the seals is the Weddell seal, which never leaves the shores of Antarctica and, indeed, prefers to remain under the coastal ice. In that watery subice world, it finds the fish and squids it feeds on; it finds warmth, for though the water may be freezing, it is considerably warmer than the subfreezing atmosphere; and it finds enclosing darkness and security.

It must, of course, breathe air, so it bites through the ice here and there to make blowholes. Occasionally it makes blowholes large enough for it to work its entire body through if it must clamber onto the ice surface.

For males this doesn't happen often. Weddell seals can dive to a depth of 600 meters (2000 feet) and, at need, remain submerged for nearly an hour, though ordinarily they come up for air every ten to thirty minutes. The female, however, must spend much more time on the ice than the male does, for only there can she bear and feed her young.

Whales and seals are the only Antarctic mammals, but there are, in addition, fifteen species of flying birds that are found in the Antarctic region. The one most at home over the continent is a predatory gull-like bird called the skua.

The most characteristic life form of the Antarctica land-mass is, however, that group of flightless birds we call penguins.

There are seventeen species of penguins altogether, all

of them native to the Southern Hemisphere. Of these, two species actually live on Antarctica. The smaller of the two is called the Adélie penguin, because it is found in Adélie Land. The Adélie penguins congregate in crowded nesting sites (rookeries) on bare ground some small distance inland. The full-grown Adélie is about 45 centimeters (16 inches) tall and weighs 6 to 7 kilograms (14 to 16 pounds).

While the Adélie penguins are on land, they are reasonably safe. There is nothing to eat on land, however, and they must dive into the ocean to catch the fish they live on. Usually, just offshore are the watching leopard seals, and the penguins must run the gauntlet if they are to eat. Overhead, too, are the circling skuas, watching for a chance to feast on penguin eggs and chicks.

The second of the Antarctica penguin species is the Emperor penguin, the largest of all living penguins. These stand over a meter (3 1/4 feet) high and weigh, at times, as much as 35 kilograms (75 pounds). —There are, however, fossils of penguins now extinct that stood 1.6 meters (5 1/4 feet) high and weighed as much as 110 kilograms (240 pounds).

The Emperor penguin rookeries are, astonishingly enough, far inland, sometimes as much as 80 to 130 kilometers (50 to 80 miles) from the shore. The Emperor penguins must walk to the rookeries and it can take them nearly a month to make the trip. Indeed, Emperor penguins are occasionally found as far as 400 kilometers (250 miles) from the nearest coast, still stubbornly trudging along—the farthest south any non-flying vertebrate is known to have reached independent of human beings.

During this walk, the penguins must fast since there is no food. Once at the rookery, the female lays her single egg and then begins the long trek back to the sea—and food.

The male, however, remains behind to incubate the egg. To do so, he places it on the upper side of his flat, webbed feet, immediately next to a bare patch of skin on his abdomen, and places a fold of feathered skin over it.

The male must keep the egg in place for some sixty

days, during which he has no choice but to continue his fast, until the female returns and takes over, by which time the egg is near to hatching. The males can then, at last, head for the sea, which they finally reach after a four-month fast during which they have lost 25 to 40 percent of their weight.

Naturally, to make this weight loss tolerable, the male Emperor penguin must gorge himself to the limit before waddling off to the rookeries in the first place. If, through misadventure, a particular male's particular consort doesn't return (no other female will do), the male, after reaching some critical point of weight loss, must abandon the egg and head for the sea. It means the growing chick must die, but if the male does not abandon it, it will die anyway and the adult with it.

When the chick hatches, the mother feeds it with food it has stored in its crop during its sea feast, but this won't last. The father must return, and for a while the parents take turns walking to the sea, gorging, and returning to feed the chick.

By the time the Antarctic summer arrives and the coastal ice begins to break up, the chicks are large enough to walk to the sea on their own, but by that time fully one fourth of the chicks have either failed to hatch or have died after hatching.

In order for an Emperor penguin chick to be large enough to make the long walk by the beginning of summer, the whole process must have begun while winter was approaching, and the male penguin must have incubated the egg through the depth of an Antarctic winter.

The males can waddle clumsily about, without dropping the eggs, and all of them huddle together in a feathered mass to endure whistling gales of up to 150 kilometers (95 miles) an hour at temperatures that go as low as −60° C. (−75° F.). The penguins within the mass are, of course, warm enough, but those at the boundaries of the huddle get the full force of the wind and constantly try to waddle inward. The whole mass of birds maintains a slow circulation that sooner or later brings every single one to the boundary for his fair share of frost.

One might assume that the Emperor penguin, as a species, was lunatic to choose this way of life; but, of course, it did not choose. Ecological pressures forced it, very slowly, into this particular niche; and adaptive change, both biological and social, kept it up.

And, believe it or not, the niche has its advantages. The very horror of the environment lends the Emperor penguin in his rookeries an almost absolute security, for no other complex form of life (barring an occasional human being or wind-blown skua) ever invades that niche.

It was about the Emperor penguin incubating its egg, that Shakespeare might truly have written:

> *Here shall he see*
> *No enemy*
> *But winter and rough weather.*

The forms of life I have mentioned are all either sea creatures or creatures that reproduce on land but must look to the sea for food. Are there any true land organisms on Antarctica, organisms that do not depend on the sea in any way?

If Antarctica had a solid ice cover, the answer would be no, since life must have liquid water to exist. Antarctica has bare patches along its coast, however. The largest bare patch is at McMurdo Sound at the eastern edge of the Ross Ice Shelf, where a stretch of ground 150 kilometers (95 miles) long and 15 to 25 kilometers (9 to 15 miles) wide is exposed. It is about as large as Rhode Island.

There are ice free "oases" in Antarctica's interior, too. Some of the mountaintops are blown free of ice and stand bare under the sky, and there are even ice-free spots here and there in the valleys. All the bare land of Antarctica put together comes to about 7500 square kilometers (3000 square miles). This is almost the area of Puerto Rico—in an ice-covered vastness which is one and a half times the size of the United States.

As a matter of curiosity, the southernmost bit of exposed land in the world is on Mount Howe and it is only 260 kilometers (160 miles) from the South Pole.

Some of the oases contain lakes, very small ones usually, which retain liquid water through the Antarctica winter, perhaps through leakage of Earth's internal heat. One such body, San Juan Pond, is about 2000 square meters (half an acre) in area and has an average depth of 0.15 meters (6 inches).

Bacteria, of course, live wherever it is in the slightest degree possible that they can. There is one species of bacteria, for instance, that is found in San Juan Pond, which, in addition to its undesirable location, is loaded with calcium chloride. (That helps it stay liquid; the addition of calcium chloride to water lowers its freezing point substantially.)

Altogether, some two hundred species of freshwater algae grow in places in Antarctica where there is exposed water. In some cases, such algae spread outward onto the nearby snow. Where there is some exposed soil, any of four hundred species of lichens and seventy-five species of moss can be found.

There are even two species of flowering plants on the Antarctic Peninsula, which stretches its thin length outside the Antarctic Circle. One of these plants is a variety of grass, the other a relative of the carnation.

Lichens have been detected on bare rock as close as 425 kilometers (260 miles) to the South Pole. That, as far as we know, is the closest land life has ever been to the South Pole independent of human activity.

Where plants exist, animal life is bound to exist, too. With the utterly insignificant plant cover of Antarctica, however, nothing larger than tiny animals can be supported. The only land animals native to Antarctica are seventy species of mites and primitive insects. The largest native land animal of the continent is a wingless fly half a centimeter (a fifth of an inch) long. One species of mite has been detected only 680 kilometers (420 miles) from the

South Pole. No other land-based animal has ever been closer to the Pole independent of human interference.

We can now return to the human invasion of Antarctica.

Up to the very end of the nineteenth century, no real landing had been made on the continent within the Antarctic Circle. John Davis (see Chapter 7) had come ashore on the Antarctic Peninsula in 1821, but his landing site was technically in the South Temperate Zone.

In the Antarctic summer of 1894–95, however, a Norwegian whaling ship, commanded by Leonard Kristensen, visited Victoria Land on the rim of the Ross Sea and there, on January 23, 1895, a party alighted and stood on Antarctica. They were the first human beings ever to stand on continental land south of the Antarctic Circle.

One of that party was the Norwegian Carsten E. Borchgrevink. He returned with an English expedition in 1898 and, with nine other men, wintered in Antarctica. It was the first time human beings ever remained on Antarctica for any extended period.

Until that winter, all Antarctic exploration had been conducted by ship. Borchgrevink put on skis, however, and set off on the first attempt to penetrate southward by land. On February 16, 1900, he attained a southern mark of latitude 78.8° S., which was farther south than any ship had ever managed to penetrate. At that point he was only 1150 kilometers (710 miles) from the South Pole.

Borchgrevink's feat fired the ambitions of others, and further sledging expeditions were planned. The obvious goal was the South Pole.

In 1902 a British explorer, Robert Falcon Scott, led a sledging expedition across the Ross Ice Shelf, and, on December 13, 1902, they reached latitude 82.28° S., only 800 kilometers (500 miles) from the South Pole. Human beings had certainly come closer to the South Pole by then than any land vertebrate had ever succeeded in doing. Scott's group had outpenguined the hardiest Emperor penguin who had ever lived.

One of Scott's colleagues, Ernest Shackleton, led another try for the South Pole in the Antarctic summer of 1908–9. On January 9, 1909, his party of four men managed to reach latitude 88.38° S., only 155 kilometers (100 miles) from the South Pole, with each man dragging his own sledge. Not even the tiniest mite had ever made its way closer than that to the South Pole. Shackleton and his men had broken the record for all forms of land life, plant or animal.

Still, they fell short of the Pole. Shackleton was forced to turn back once it was clear that to travel farther would mean that the food supply would not last the return journey.

All was now set for the final push. Two candidates were in the field. One was Scott again, and the other was the Norwegian explorer Roald Amundsen.

Amundsen concentrated on dogs. He set off on October 20, 1911, with fifty-two dogs pulling his sledges. This was a shrewd move for he did not need to carry much food for his dogs. They were carnivorous, after all, and as he proceeded, he killed and fed the weaker dogs to the stronger ones. Between that and the fact that the dog-drawn sledges would move faster than human-drawn ones, there was no danger of being forced back by a shortage of the food supply.

Amundsen reached the South Pole on December 14, 1911, left his marker there, and then began the return trip. It would have been dangerous to linger. He hastened back for the coast and safety, reaching it on January 21, 1912. He still had twelve dogs surviving and food left over. His expedition had suffered not one human casualty.

Scott planned his expedition differently. He had fewer dogs, but he had ponies in addition and motorized sledges as well. He started inland on October 24, 1911, four days after Amundsen.

Motorized sledges would have been perfect if they had worked, but this was 1911 and the state of the art had not moved far forward. All the motors broke down in fairly short order.

Then, too, the ponies were a miscalculation. They were herbivores and fodder had to be carried for them. They had to be shot when there was still nearly 810 kilometers (500 miles) to go because there was no more food for them. The dogs might have been fed on horsemeat, but there weren't enough of horses to serve the purpose. They were sent back with some of the men. The last 650 kilometers (400 miles) had to be traversed by man-drawn sledges.

Finally, Scott and four companions reached the South Pole, on January 17, 1912, and found Amundsen's marker there. He had been there five weeks earlier.

The five men had to find their way back to the coast, with no animals and with their food supply dangerously low. One man died soon after the return journey started and a second, L. E. G. Oates, feeling he was weakening, deliberately walked off into the snow and cold to die so as to be no burden on the others.

The last three might have made it but a blizzard struck when they were only 18 kilometers (11 miles) from safety and pinned them down in their tent. Day after day, they made ready for the final lunge and day after day the blizzard monotonously continued. It lasted nine days and in the course of that the three men died of hunger and exposure on or about March 29, 1912. With true British understatement, Scott's last diary entry read, "It seems a pity, but I do not think I can write anymore."

Yet if human beings had finally reached the South Pole, we, as a species, were, like Scott, only second to the pole.

To be sure, human beings (and their dogs and parasites) were the first land organisms to stand at the pole, but not all organisms are tied to the land.

The skuas, skimming through Antarctica's airspace, range all over the continent and it seems inevitable that every once in a while one of the birds has flown over the South Pole. They are the only species of living creatures that have reached the South Pole independently of man and there is no question, obviously, that they have reached it first.

Great as the human accomplishment is, we have, in this respect, been but second to the skua.

But wait. We are talking about Earth as it now is. The South Pole has always been where it now is, relative to the Sun (if we allow for the regular changes of the precession of the equinoxes, nutation, and so on) for perhaps 4 billion years at least.

Antarctica, however, hasn't.

The Earth's crust is cracked into about a dozen good-sized plates. Driven by some internal engine (perhaps the slow circulation of material within the Earth's mantle, itself powered by Earth's internal heat), these plates shift position. They pull apart at some joints, while magma wells up to form volcanic areas. They crumple together at other joints, forming mountain ranges—or else one plate may dive under another to form oceanic trenches.

And on the plates are the chunks of continental granite, riding high on a bed of sea floor basalt. Slowly the continents approach each other and recede, and every once in a while they come together in such a fashion as to form a single supercontinent called "Pangaea" (Greek for "all-Earth").

Some 225,000,000 years ago, the most recent Pangaea existed, and ocean water rolled over both poles. What was then the Antarctic Ocean was ice-covered as the Arctic Ocean is now, and, undoubtedly, under the ice there was a varied sea life in existence so that myriads of life-forms preceded both skua and human beings at the South Pole.

But Pangaea broke up, and its portions, on different plates, moved apart ("continental drift"). About 40,-000,000 years ago, one fragment of Pangaea broke up into Madagascar, Australia, India, and Antarctica. India veered northward and finally collided with Asia to form the great Himalayan mountain range at the crumpling line of collision. Antarctica moved southward for a rendezvous with its frozen destiny.

For millions of years, though, before Antarctica moved through the Antarctic Circle and over the South Pole, the continent had a mild climate. In the days when amphibians ruled the land and early reptiles were beginning to appear, it must have teemed with life.

Scott, himself, the tragic second human at the South Pole, had come upon a deposit of coal in Victoria Land in 1903—and where there is now coal, there was once copious plant life. This, in itself, proved that the Antarctic was warm in times past or that Antarctica wasn't always in the Antarctic. For half a century, it was the first guess that was the popular one, but for the last twenty years, we are convinced that it is the second that is correct.

Nor could the coal have originated in some way not involving life. Fossilized trunks of trees have been found and imprints of leaves on rocks. The prints are detailed enough to be identified as having been formed by leaves of *Glossopteris,* a plant that flourished in the tropical jungles of Africa and South America 225,000,000 years ago.

Where plants exist, animal life is sure to exist as well, but Antarctica is not exactly a happy hunting ground for paleontologists. Ice, kilometers thick, covers the ground where fossils might be found—but not absolutely everywhere.

In December 1967 a New Zealand geologist, Peter J. Barrett, came across something on Graphite Peak that looked like a pebble but turned out to be a fragment of bone that was eventually identified as a part of the skull of an ancient amphibian called a "labyrinthodont." Others began to comb the area, and in March 1968 the American paleontologist Edwin H. Colbert discovered the lower jawbone of a labyrinthodont in a cliff about 520 kilometers (325 miles) from the South Pole. The jawbone was surrounded by fossils of swamp plants.

The jawbone was quite like the labyrinthodont relics located in Africa, Madagascar, and Australia, and the labyrinthodonts were fresh-water creatures who could not have crossed oceans. Their existence in all these places showed that the land must once have been a single piece and offers the best proof that continental drift had actually taken place.

In 1969 fossil fragments of a small hippopotamuslike reptile called *Lystrosaurus* were discovered at Coalsack Bluff, about 650 kilometers (400 miles) from the South Pole. Then, on November 10, 1970, James Colinson discovered the first complete vertebrate fossil ever found in Antarctica, a foot-long cynodont, a mammallike reptile.

It seems quite obvious, then, that however desolate Antarctica is now, it was once rich with life. If we could dig straight down from the South Pole, that spot reached, after so much effort, by human beings, we would find that any number of creatures had once been on that spot and left their fossils behind and that they had long antedated skua and human being alike.

But those long-dead creatures had never gotten to the South Pole under their own power.

Antarctica itself, crawling with infraglacial slowness, had brought them there.

D PLANETS

9. The Sons of Mars Revisited

I dined at a very pleasant French restaurant last night and, at the conclusion of the meal, when the *maître de* arrived to express his hope that we had had a pleasant dinner, our hostess assured him we had indeed had one and said, "Do you know who our guest is?"

Then she turned to me and said, "Do you mind?"

Of course, I minded. It's not that I object to the notoriety, but too many of those who, for one reason or another, are enthusiastic about my writing overestimate my status. They seem to think I'm a household word, and I'm not.

It's my experience that a very large majority of Americans, and a fairly substantial majority of even intelligent Americans, have never heard of me and, after my importance has been explained to them, prefer to continue not hearing of me.

I suspected, therefore, that my hostess would be embarrassed and that I would be made uncomfortable. It was too late to stop her, however, so I nodded my resigned permission.

My hostess said enthusiastically, "Our guest is the famous writer Isaac Asimov."

The *maître de,* schooled to every politeness, smiled very patiently and said, "I am pleased to meet you." It was, however, quite obvious to me that he hadn't heard of me, that he didn't care that he hadn't, and that he wasn't particularly pleased to meet me.

What saved the situation was that ten feet away stood the pretty waitress who had been dancing attendance on all our wants for two hours and with whom I had been exchanging badinage (as is my wont). At the mention of my name, she suddenly gasped.

To ignore the *maître de* was, for me, the work of a moment; to smile broadly at the pretty waitress and say, "Have you read my books?" was the work of another.

"Oh, yes," she said, "but I never thought you really existed. I thought you were at least five men."

"The usual reaction of women, my dear," I said, suavely, and the whole evening ended on a high point.

So watch out for the unexpected side effects, as in the case of Mariner 9—

On May 30, 1971, Mariner 9 was launched. The intention was to have it move into orbit around Mars and photograph its entire surface. There had been some thought of having it also photograph the two Martian satellites, but whoever decided on priorities insisted that the probe was to concentrate on Mars only. The scientists in charge, however, arranged a certain flexibility that would allow some study of the satellites if the possibility turned up.

The possibility did indeed turn up. When Mariner 9 arrived at Mars, it found a major planetary dust storm had shrouded the planet in an impenetrable cloud and there was nothing for Mariner 9 to do but circle the planet and twiddle its thumbs. Rather than waste its time altogether, it was ordered to take a look at the satellites, and this unexpected side effect of the dust storm produced excellent results.

Later probes, Viking 1 and Viking 2, also studied the satellites of Mars, so that we know infinitely more about them now than we did a decade ago.

We did know something about them, of course, from the moment they were discovered by the American astronomer Asaph Hall, in 1877. He named them, very appropriately, Phobos (fear) and Deimos (terror) after the sons of Mars (or, more correctly, Ares) in the Greek myths.

The distance of these satellites from the planet and their

period of revolution were worked out at once, and I wrote about this and how one could deduce from it the appearance of the satellites in the Martian sky in my essay "Kaleidoscope in the Sky" which appeared in the Doubleday collection *Science, Numbers and I*.

What we still didn't know at that time was the exact size of the satellites. For that we had to wait for the Mars probes, and it is to discuss this that I now revisit the sons of Mars.

From the very start it was certain that the Martian satellites were very small. Even when they were at their closest to us, only 56,000,000 kilometers (35,000,000 miles) away, they remained very dim.

Satellites such as Ganymede (Jupiter) and Titan (Saturn) are brighter than Phobos or Deimos even though the former pair is much farther away. Though Triton, the satellite of Neptune, is never closer to us than 4,350,000,000 kilometers (2,700,000,000 miles), which is 75 times as far from us as are Phobos or Deimos at their closest, that distant satellite is nevertheless almost as bright as Phobos or Deimos. This, despite the fact that Triton is lit only by what is, with reference to itself, only a very dim and distant Sun.

From this alone we know that the Martian satellites must be excessively tiny. Why else should it have taken so long to discover them?

But how tiny is "tiny"? Astronomers couldn't tell. If the satellites would only show a visible disk that could be measured, then, from their known distance from us, their diameters could be worked out. They were too small to show a visible disk, however, even under the greatest magnification.

Failing that, an idea of the diameter of a satellite could be obtained from the amount of light it caught from the Sun and reflected to us. We knew how far the satellites were from the Sun and from us, so given the amount of light we actually receive, we could calculate how much they receive per square kilometer of surface, if those square kilometers reflected every bit of light that fell on them.

However, any object reflects only a fraction of the light that falls on it (that fraction is its "albedo") and we don't know what the albedo of Mars's satellites is. If we knew the albedo, we would at once know the satellite size.

We could, of course, make some logical assumptions. For one thing, the Martian satellites are far too small and far too close to the Sun to have any trace of atmosphere or water. Their surfaces would have to be bare rock.

Our Moon's surface is bare rock and that reflects very little light. Despite its apparent brightness, the Moon has an albedo of only 0.06; it reflects only about 1/16 of the light that falls on it.

In 1956 the Dutch-American astronomer Gerard Peter Kuiper assumed that Phobos and Deimos had the albedo of the Moon. Based on this assumption, he estimated that Phobos was about 12 kilometers (7.5 miles) in diameter and that Deimos was about 6 kilometers (3.75 miles) in diameter.

This was the best that could be done until 1969. In that year, the Mars probe Mariner 7 passed by Mars and took photographs of its surface as it passed. One of these photographs happened to be taken just as Phobos passed in front of the lens and the satellite showed up as a black silhouette against Mars.

The study of that photograph showed that Phobos was unexpectedly dark. It reflected an even smaller fraction of the light that fell upon it than the Moon did. This meant that it would have to be somewhat larger than Kuiper had estimated it to be, if it were to send us the light it does. The same was likely true of Deimos, if it, too, had a lower albedo than the Moon.

The second piece of information was that the silhouette of Phobos was not round but oval.

This was not, actually, very surprising.

The Sun and Moon are round in outline when viewed from any vantage point, but Earth and Mars, when viewed from above the equatorial plane, appear slightly elliptical.

Saturn and Jupiter, when viewed so, appear pronouncedly elliptical.

These elliptical appearances are the result of rotation. If a heavenly body did not rotate, or rotated only very slowly, then, when viewed from a distance at any angle, it would appear round—provided it was large enough.

A large body, the size of the Moon or larger, produces a gravitational field intense enough to compact all its matter as close to the center as possible and this automatically produces a sphere. The sphere may not be perfect, for there can be surface irregularities. These irregularities are very small compared to the total size of the sphere so that if the Earth were reduced to the size of a billiard ball, it would be smoother than a billiard ball despite all its mountains and valleys.

As a body with less and less mass is considered, its gravitational field is less and less able to compact it, and the surface irregularities, though no larger in an absolute sense than those on larger worlds, become much larger when viewed as fractions of the total diameter. In that case, the world has a sensibly irregular shape.

This was already known of the asteroid Eros, for instance. It is larger than Phobos and Deimos and delivers light that waxes and wanes regularly in intensity. The assumption is that, as it revolves, it presents cross sections of different size. It could be roughly brick-shaped, for instance, and when we view it broad side on, we would receive more light from it than when we view it narrow side on.

Why, then, shouldn't Phobos and Deimos be irregularly shaped as well?

They are. Mariner 9's photographs showed the satellites to be remarkably potato-shaped, right down to the "eyes" which were, actually, craters.

If you smoothed out, in imagination, the crater irregularities of the two satellites, you found that the average shape was an asymmetric ellipsoid which had to be defined by three measurements.

For instance, of all the lines through the center of either satellite, one line, stretching from surface to surface, would be the "longest diameter."

Suppose you draw the various lines through the center which are perpendicular to the longest diameters. Of these, one is shortest and is the "shortest diameter." The diameter which is at right angles to both the longest and shortest diameters is the "intermediate diameter."

For Phobos, the longest diameter is 28 kilometers (17 miles), the intermediate diameter is 23 kilometers (14 miles), and the shortest diameter is 20 kilometers (12 miles). Even the shortest diameter is distinctly larger than Kuiper's estimate of the average diameter of Phobos.

For Deimos, the longest diameter is 16 kilometers (10 miles), the intermediate diameter is 12 kilometers (7.5 miles) and the shortest diameter is 10 kilometers (6 miles). Again, this is distinctly larger than the Kuiper estimate.

Even so, these are small worlds. If you imagine Phobos placed on New York City in such a way as to cover as much ground as possible, it would cover the boroughs of Brooklyn and Queens. Deimos, placed similarly, would cover the borough of Brooklyn alone.

The satellites are not too small, however, to support a comfortable astronautical base in times to come.

The surface area of Phobos is 1500 square kilometers (570 square miles), which is a little over half that of the state of Rhode Island. The surface area of Deimos is 400 square kilometers (160 square miles) which is a little over half that of the city of New York.

As for volume, Phobos contains about 5400 cubic kilometers (1300 cubic miles) while Deimos contains 780 cubic kilometers (190 cubic miles). Phobos is thus about 7 times as voluminous as Deimos.

If our Moon were hollow, it would take over 4,000,000 bodies the size of Phobos, tightly packed, to fill it. It would take 28,000,000 bodies the size of Deimos to do the same job.

The best estimates, at the moment, are that Phobos and

Deimos have densities of 2 grams per cubic centimeter (125 pounds per cubic foot). If so, then the mass of Phobos is 10,800,000 billion kilograms (12,000 billion tons) and of Deimos 1,500,000 billion kilograms (1700 billion tons).

From the mass of the satellites and the diameters, we can calculate the value of their surface gravity as compared to that of Earth.

The average surface gravity on Phobos would be 0.0007 that of the Earth. It would be a little higher at the end of the shortest diameter, a little lower at the end of the longest diameter, but not enough to be noticeable to any astronaut using the satellites as bases.

A person weighing 70 kilograms (154 pounds) on Earth would weigh about 50 grams (1.7 ounces) on Phobos. He would weigh half of that on Deimos.

Both Phobos and Deimos revolve about Mars in such a way as to show the same face to Mars at all times, just as the Moon always shows the same face to Earth as it revolves. Each of the Martian satellites turns so that one end of its longest diameter points always toward Mars. (The other end, naturally, points always away from Mars.)

This means that Phobos and Deimos each makes one turn relative to the stars as it moves once around Mars. The period of revolution of each is therefore also its "sidereal day" ("sidereal" is from a Latin word meaning "star"). The sidereal day of Phobos is 7.65 hours and of Deimos 30.30 hours.

Each time Phobos or Deimos has moved once about Mars, however, it has followed Mars in the revolution of the latter about the Sun. This produces a small apparent motion of the Sun in the reverse direction, so that the Sun seems to complete its circuit of the sky, as seen from Phobos or Deimos, more slowly than the stars do.* (This is true on Earth also. The Sun moves once about our sky

* See "The Dance of the Sun," in the Doubleday collection *The Solar System and Back*.

in 24 hours, but each star makes a complete circle in only 23 hours and 56 minutes.)

The period from sunrise to sunrise ("solar day") on Phobos is 0.0036 hours (20 seconds) longer than the period from star rise to star rise. The extra length is greater on Deimos, which takes a longer time to make its revolution, thus giving the Sun a longer time to drift backward. The period from sunrise to sunrise on Deimos is 0.056 hours (3.3 minutes) longer than the period from star rise to star rise.

Suppose, now, that you are standing on the end of the longest diameter of Phobos, the end facing away from Mars. That end would always face away from Mars and you would never see the planet you were circling. On the other hand, you would see the stars wheeling about the sky from east to west, as you do on Earth, but moving a little over three times as quickly.

The Sun would also rise in the east and set in the west but wouldn't quite keep up with the stars. Each sunrise would be 20 seconds later, relative to the stars, and the Sun would therefore lose one complete circuit of the sky in 687 days, which is the length of the Martian year.

If you are on the end of the longest diameter of Deimos, the end facing away from Mars, you would see the stars moving from east to west at a pace just a little slower than would be the case on Earth. Each Sunrise would be 3.3 minutes further behind the stars, but since Deimos makes fewer rotations in the course of the Martian year, this, too, would amount to a total lag of one day in the course of that Martian year.

The circumference of Phobos from one end of the longest diameter to the other and back is 79 kilometers (49 miles). For Deimos, the corresponding figure is 44 kilometers (27 miles). To walk the maximum circumference of Phobos, then, is like walking from New York City to Trenton, New Jersey. To walk around Deimos is like walking from Fort Worth, Texas, to Dallas, Texas. To put it another way, any two points on Phobos' surface are separated by less than 40 kilometers (25 miles) and on Deimos by less than 22 kilometers (14 miles).

Given the circumference of the satellites and the length of the day, it appears that Phobos is turning, with respect to the stars, at a rate of 10.3 kilometers (6.4 miles) per hour and Deimos at a rate of 1.5 kilometers (0.9 miles) per hour. The rate is very slightly slower with respect to the Sun.

This means that if we were to trot westward at a brisk, but not impossible, speed on Phobos, we could keep up with the turning sky. If the Sun were not in the sky, it would never rise for us as long as we managed to keep running. Or, if it were overhead, it would never stop being overhead as long as we managed to keep running.

On Deimos, a leisurely stroll would do the same. In fact, if we walked briskly westward on Deimos, we would overtake the Sun and see it rise in the west.

And what about Mars as seen from the satellites? As long as we remain at the far end of the longest diameter (or near it), Mars is never in the sky. If, however, we begin at the far end and move away in any direction without changing that direction, then in some 20 kilometers (12 miles) on Phobos or 11 kilometers (6.8 miles) on Deimos we will see Mars looming up above the horizon.

By the time we reach the opposite end of the longest diameter, the one facing Mars, then Mars will be directly overhead and it will stay there as long as we remain on that point on the satellite.

In moving from the far end of the longest diameter to the near end, the appearance of Mars and its lifting above the horizon must be a beautiful and awe-inspiring sight.

Not only is Mars larger than our Moon, but it is much closer to its satellites than the Moon is to Earth. It bulks much larger, therefore, in the satellite sky than the Moon does in ours.

As seen from Deimos, Mars is a little over 30 times as wide as the Moon appears to us, and its apparent area is just about 1000 times as great as the Moon's apparent area is to us.

As seen from Phobos, which is considerably nearer to Mars than Deimos is, Mars is 78 times as wide as the Moon appears to us and has 6100 times the apparent area.

Mars appears so huge as seen from Phobos that four globes the size of Mars, placed side by side, would stretch nearly from horizon to horizon, while nine would encircle the sky.

Mars reflects a larger fraction of the light falling upon it than the Moon does, but Mars is farther from the Sun than the Moon is and gets less light per square kilometer for that reason. Taking all these factors into account, Mars, as seen from Deimos, is, at its brightest, 940 times as bright as the full Moon appears to us. Mars, as seen from Phobos, is, at its brightest, 5700 times as bright as our full Moon.

It might be thought that if Mars is so much brighter in the satellite skies than the Moon is in ours, then Mars must really be the dominating body of those skies. So it is, in terms of sheer size, since the fact that it is so much larger than the Moon means that it is so much larger than the Sun (as seen in our sky) as well. It is even larger if we compare it to the Sun as seen in the satellite skies, for out there the Sun is only 5/8 as wide as it appears to us and is only 2/5 as bright.

Even so, however, as seen from Deimos, the shrunken Sun is still 212 times as bright as Mars ever gets to be. And on Phobos, where Mars is so bloated, the Sun is nevertheless 35 times as bright as Mars ever gets to be.

What's more, when the Sun is in the satellite skies, Mars simply isn't at its brightest. Not only does the Sun's greater brightness wash it out, but with the Sun in the satellite skies, less than half of Mars's surface is illuminated.

As the Sun circles the sky, Mars goes through a series of phases just as the Moon does as seen from the Earth; it goes through a complete series of phases once each revolution of the satellites.

Suppose we are standing on Phobos at the end of the longest diameter, the end facing Mars. Mars is at zenith, huge and bright, but it is just about half full, for it is sunrise and the Sun's light is coming from the east, so that the western half of Mars, as seen from Phobos, is dark.

As the Sun rises rapidly, the lighted portion of Mars shrinks, and becomes an ever-narrowing crescent. One and a half hours after sunrise, the Sun reaches Mars, slips behind it, and is eclipsed. Mars is now totally dark, for only its far side is lit by the Sun.

Well, not *totally* dark. The Sun illuminates the Martian atmosphere and against the black of the sky, a luminous ring outlines the darkened body of the planet.

The Martian atmosphere contains the pink dust of Mars so that the outlining circle is a delicate pink and undoubtedly very beautiful.

It is not a symmetrical pink circle. When the Sun slips behind the eastern edge of Mars, the eastern rim of the atmosphere is illuminated more brightly than the western rim. As the Sun moves farther westward, however, the brightness of the eastern rim dims and that of the western rim brightens until, after twenty-five minutes, the circle of pink light is more or less evenly bright. Then, the western rim brightens further, while the eastern rim dims further until, after another twenty-five minutes, the Sun's flame bursts out from behind the western edge of Mars.

As the Sun sinks, Mars becomes an ever-thickening crescent on its western side. By sunset, which comes an hour and a half after the Sun has emerged from behind Mars, Mars is half full again; its western half being lighted this time and its eastern half dark.

After the Sun sinks behind the western horizon, the lighted portion of Mars's globe expands further. Just under two hours after Sunset, the Sun is shining directly down on the other end of the longest diameter, so that its light slips past Phobos and onto the side of Mars facing the satellite. Mars is now at its full and brightest—all the more so because there is no Sun in the sky to compete.

But the western edge of Mars begins to darken at once, and Mars shrinks. In just under two more hours, it is in the half phase and the Sun is rising again.

For Deimos, the cycle is much the same but slower. It takes four times as long for the Sun to go through each of these stages.

Mars, as seen from Deimos, is smaller than as seen from Phobos. The Sun, however, as it moves behind the smaller Mars, does so at a slower pace, so that it remains behind Mars for eighty-two minutes.

In "Kaleidoscope in the Sky" I described the size of the satellites as seen in the Martian sky, making use of Kuiper's estimate of their size. Let me, in this revisit, give you better figures.

When Phobos is seen directly overhead by an observer standing on the Martian equator, it is then at its maximum apparent size. We see its longest diameter facing us, so that we see it with its intermediate diameter in some particular direction and the shortest diameter at right angles to that. Phobos would there appear as an oval body 13.3 by 11.2 minutes of arc. Even its widest diameter would be less than half that of the Moon as seen from Earth, and Phobos' total apparent area would at best be only 1/6 that of the Moon.

As for Deimos, it would appear only as a fat star, 2.0 by 1.6 minutes of arc, with an area less than 1/300 of the Moon.

Phobos would never seem more than 1/16 as bright as the full Moon in our sky, since it is bathed in weaker sunlight and has a lower albedo than the Moon. Deimos would never be more than 1/1000 as bright as the full Moon in our sky.

Nevertheless, the three brightest objects in Mars's sky would be the Sun, Phobos, and Deimos, in that order. Fourth would be (at its brightest) Earth!

10. Dark and Bright

As is well known to everyone but people, Harlan Ellison and I love each other and there is no feud between us.

Of course, Harlan has the fastest quip this side of Dorothy Parker and very few people can stay undemolished for more than fourteen seconds (at the outside) if they start up with him. Fortunately, I am an exception and have frequently lasted an easy thirty-seven seconds, so we have at each other now and then for the fun of it and that is what starts the rumors of feuds.

Naturally, in so doing, we take off on each other's physical characteristics, since neither of us is exactly well known as models of good taste in the heat of quippery. Since Harlan is slightly below the American average in height and I am slightly above the American average in weight, when we get together I make short jokes and he makes fat jokes.

We were together at the Nebula Awards dinner on Saturday, April 30, 1977* and I made some casual mention of planning to lose some weight.

Whereupon Harlan said, "In that case, I'll gain some height."

Well, maybe Harlan can't do that, but it takes us back to the Martian satellites, which we were discussing in the previous chapter and which seemed to have gained in size

---

* I am not even going to tell you who won the Nebula in the novelette category on that great occasion, because I am incredibly modest.

as we learned more about them—with interesting results that we will now continue to probe.

As I said in the previous chapter, the first good look at Phobos, the inner satellite, showed us that it was cratered. In fact, both satellites, Phobos and Deimos, are crater-saturated, in the sense that every spot on their surface is part of one crater or another.

This is not really a surprise. There is a general consensus among astronomers now that the various bodies of the solar system were formed by accretion—that is, small particles conglomerated into larger particles, which in turn conglomerated into still larger ones, and so on.

Finally, when sizable bodies were formed, they were bombarded by slightly less sizable bodies, later collisions forming craters that demolished and masked those formed by earlier collisions, until, finally, the last collisions left craters that remain, more or less, to this day.

Perhaps 4 billion years ago the chief period of crater formation was over. Space was cleared of most of the sizable free pieces and the solar system took on the shape, more or less, that it now possesses. There were still occasional collisions and now and then a new crater, even a spectacular one, might be formed, but they represented mere footnotes. The book itself was written.

Once all the final craters are formed, is there any force that would wipe them out?

Sure. If a body is large enough to ignite as a star and if it becomes a glowing mass of gas, any craters it might have had in an earlier solid stage would go. So would all irregularities except for gouts and swirls of incandescent gas. The Sun is an example of that.

Second, if a body has a sizable ocean, or a thick atmosphere, or heavy volcanic activity, or an active biological system, or any combination of these things, craters might be reduced or even be obliterated by the erosive effects of water and wind, by the efflux of molten lava, by the working of living things, or by any combination of these.

The best example of that is Earth, where all four effects are to be found and where virtually none of the primordial craters, or even very recent ones, geologically speaking, still exist. Meteor Crater in Arizona is the best example of one that exists. Its nature is unmistakable because it was only formed a few tens of thousands of years ago in a dry area. There are also a few other circular formations on Earth, usually partly water-filled, which can be recognized as one-time craters from the air, and that's all.

We can't tell what goes on under the heavy atmospheres of the four outer giant planets, but it doesn't seem conceivable, if anywhere below the visible cloud layers there is a solid surface, that heat and atmospheric action would not have wiped out any craters that might have once existed.

We know a little more of what goes on under the cloud layer of Venus, which is closer, smaller, and less gaseous than the outer giants, thanks to radar-reflection studies. Recently, astronomers have detected signs of something on Venus that may indicate a volcano even larger than Olympus Mons† on Mars.

Assuming ample volcanic action on a planet almost as large as Earth and even hotter, and remembering that Venus has an atmosphere about a hundred times as dense as ours, we can be reasonably sure that what craters existed there, have in large measure been eroded and melted away. Still, since Venus lacks an ocean or (almost certainly) life, it may be more cratered than Earth is.

Even on Mars, which is colder and smaller than either Earth or Venus, which has only a trifle of atmosphere, and which seems to have no life (except just possibly some microscopic forms), there has been enough volcanic action to wipe out the primordial craters over about half its surface.

Any bodies smaller than Mars, however, which are quite likely to have none of the attributes required for crater erasal, should be heavily cratered.

This seems to be true of Mercury, for instance. Most of

† See "The Olympian Snows," in my Doubleday collection *The Planet That Wasn't.*

it has been mapped by Mariner 10 and it seems uniformly stippled with craters. It is certainly true of the Moon—except that there are some areas that seem to have been leveled by lava flows to form the "maria" (seas). Perhaps this is the result of particularly large collisions at the end of the crater-forming process, but just about all the maria seem to exist on the side of the Moon that perpetually faces the Earth, which is rather a puzzle.

It is probably true of all the satellites of the outer planets, except possibly for Saturn's Titan, which has a respectably thick atmosphere.‡

But Phobos and Deimos—which are tiny, airless, water-less, lifeless and heatless? What can possibly erase their craters? Nothing.

They are the closest objects to Earth (except for an occasional flit-by asteroid) which possess surfaces that may be exactly as they were when the crater-forming period ceased. They would therefore represent the oldest surfaces one can possibly have in the solar system and for that reason alone it is worth examining the Martian satellites as closely as possible and even planning a human landing on them some day.

Based on the photographs of Phobos by Mariner 9, a preliminary map was made showing about fifty craters. On this map the satellite's equator and the prime meridian cross at the end of the longest axis—that end which faces Mars (see Chapter 9).

One large Phoban crater, about 6 kilometers (3.7 miles) across, near the south pole, is named Hall. That is appropriate since it was Asaph Hall who discovered the satellites.

An even larger crater, at the equator and just a little west of the prime meridian, is named Stickney. This is even more appropriate since when Hall had decided to give up the search, his wife, whose maiden name was Angelina Stickney, urged him to try one more night—and

‡ See "Titanic Surprise," in my Doubleday collection *The Planet That Wasn't.*

that proved the crucial night of discovery. Stickney is 10 kilometers (6 miles) across and is about 40 percent as wide as the maximum diameter of Phobos.

Since the map was made, new and even better observations by the Viking probes have revealed narrow parallel grooves along Phobos' cratered surface, those grooves being more or less parallel to the equator. The significance of these is as yet unknown.

Deimos, a smaller body, has, as one might expect, smaller craters. The largest, which is 2 kilometers (1.3 miles) wide is named Voltaire. The next largest, which is half as wide, is named Swift.

Why those two names? Well, thereby hangs a tale—

As soon as Jupiter was found to have four satellites (that was in 1610), the German astronomer Johannes Kepler, who loved to play with numbers, pointed out that Mars ought to have two satellites. After all, Earth had one and Jupiter had four, so wasn't it only natural that Mars, which was the planet in between, should have two?

Actually, there is nothing to the argument, for the number of satellites a planet may have is indefinite. For instance, the current number of known satellites of Jupiter is not four, but fourteen.

Just the same, there is always an attraction to neatness in numbers and Mars's two imaginary satellites cropped up now and then in literature, especially since someone as respected as Kepler had mentioned it.

In 1752, for instance, the French writer Voltaire wrote a book called *Micromégas,* in which giant beings from Saturn and from the star Sirius visit Earth to observe humanity and to marvel at its follies. On their voyage to Earth, Voltaire has them observe Mars's two satellites in passing.

Even earlier than that, in 1726, the English writer Jonathan Swift published *Gulliver's Travels.* In the third part of that book, Swift has Gulliver visit the mythical land of Laputa, where the astronomers were supposed to possess advanced telescopes that made it possible for them to discover the two Martian satellites.

Swift's discussion of the two satellites was particularly important because he actually gave his imaginary worlds specific distances from Mars and special periods of rotation. He placed the inner satellite about 20,000 kilometers (12,500 miles) from Mars, with a period of revolution of 10 hours; and the outer satellite about 34,000 kilometers (21,000 miles) from Mars, with a period of revolution of 21 1/2 hours.

That's not very far from the truth. The actual inner satellite, Phobos, is only half the distance of Swift's and the actual outer satellite, Deimos, is only two thirds the distance of Swift's, but that still leaves it a remarkable bit of intuition.

How did Swift know? Mystics have tried to make much of this by supposing Swift to have had access to hidden knowledge or to have had some sort of paranormal powers. (Some have even, in joke, suggested he might have been a Martian.)

Actually, there is no secret to Swift's good guess; he was using his head. As for having Mars possess two satellites, that number was in the air, as I said, and the fact that it turned out to be true in the end was pure coincidence. From then on, though, it was just deduction.

Mars is so close to us that if it had a good-sized satellite at a reasonable distance from itself, those would have been discovered by Swift's time. After all, Jupiter and Saturn are much more distant than Mars is and yet satellites had already been discovered for those planets. The conclusion was that if Mars did have two satellites, they would have to be quite small so as to be hard to see, or quite close to the planet so that they were lost in Mars's glare of light, or both.

For that reason, Swift, who was no dummy, put them close to the planet. Then, if they were close, they would have to revolve about Mars in short periods. In fact, from the distance, the period of revolution can be calculated by Kepler's third law and Swift refers to that law in the passage.

In any case, the craters on Deimos are named appropriately, though I would have, given my choice, reversed matters and given Swift's name to the larger.

What are the satellites doing there? In general, there are three ways of explaining the existence of satellites—

1. The coalescing fragments of a planetary cloud conglomerate into a single body which then splits off into one or more fragments.

2. The coalescing fragments form more than one nucleus to begin with, the largest one at the center and smaller ones at the outskirts of the original cloud.

3. The coalescing fragments form one nucleus which then proceeds to capture small bodies that happen to pass close by under conditions where gravitational capture is possible.

The first possibility was very popular in connection with the Earth and Moon at one time but has now been largely given up. It seems to be growing less and less popular in general, so let's forget it.

That leaves the other two possibilities. They are the simultaneous-formation theory and the capture theory— but how does one decide between the two?

An easy way of deciding would be to suppose that if a planet and a particular satellite were formed out of the same cloud, they would have the same general composition; while if they were formed out of different clouds, they would have different, even radically different, compositions.

It's not quite as easy as that, however.

A particular whirling cloud out of which a planet is forming may not have the same chemical composition throughout. There may be a settling-out process taking place, with denser materials settling toward the center faster than less dense materials would. Thus, the four major satellites of Jupiter have densities that seem to decrease steadily with distance from Jupiter, but no one thinks they were captured by Jupiter for that reason.

Then, the particularly large central planet would have a more intense gravitational field than small satellites would.

The planet would attract the light gaseous elements that the satellites would not. Thus, Jupiter is largely hydrogen and its four major satellites are composed largely of ices and rocks, but no one suspects them of having been captured for that reason.

In the case of Earth and Moon, there is a much smaller disparity in size than in the case of Jupiter and its satellites, so that we would expect lesser chemical differences to exist if Earth and Moon were formed out of the same cloud of material.

The Moon's chemical structure is different from that of the Earth, but is it different enough to indicate formation out of different clouds? Opinions differ. The Moon's peculiar chemical structure may arise out of the fact that because of its smaller mass and lesser gravity, it never had air or water and was therefore more exposed to changes in temperature. On the other hand, the common occurrence of glassy material on its surface may show that it was at one time much closer to the Sun than Earth was.

The evidence is, so far, equivocal, and the astronomers find, in frustration, that even a man landing on the Moon and bringing back Moon rocks for them to study in detail has not forced a clear decision between the simultaneous-formation theory and the capture theory of the Moon's origin.

But what about Mars and its satellites? Can we come to any decision there?

What about simultaneous formation?

Why not? Might not the Martian satellites be two final pieces of conglomerating matter that just happened to be moving in the right direction and at the right speed to take up an orbit about Mars and never crash into it to form the final two craters?

It could be that there were a great many such satellites 4 billion years ago which collided with each other, bouncing back and forth, till all hit Mars, leaving a final pair which just happened to be far enough apart not to collide with each other ever, so that neither had anything to stir it out of its safe orbit.

But in that case, how is it that they are both moving in orbits about Mars that are just about circular and are just about in the plane of Mars's equator? Why might not the two last pieces just happen to have elongated orbits traveling about Mars in any plane?

Apparently, such satellites are subjected to a tidal drag which tends to move them into those orbits where the drag is at a minimum—which would be a circular orbit in the equatorial plane. Such tidal drag is more effective, the more massive the central planet and the less massive the satellite, and even in a circular orbit in the equatorial plane a small satellite might be pulled closer to the planet through tidal drag.

Some years ago it seemed that Phobos was approaching Mars at a rate that was out of proportion to what one would expect from its size. The rate of approach could only be explained by supposing Phobos to have an astonishingly small mass.

The modern Soviet astronomer Iosif Samuilovich Shklovsky suggested that Phobos couldn't possibly be that unmassive unless it were hollow and he even postulated that it might be a space station built by an advanced Martian civilization. (That was, of course, at a time when there still seemed to be some vague possibility that there were canals on Mars built by intelligent beings.)

The notion of "Space-station Phobos" was destroyed in three ways. In the first place, it became clear that there was no advanced civilization on Mars, thanks to the Mars probes. Second, the first photographs of Phobos showed it had to be solid since anything capable of forming the craters upon it would have smashed it into fragments if it had been hollow. Third, the rate at which Phobos was approaching Mars turned out, on more careful measurement, to be no faster than would be expected for a solid body of Phobos' size.

Even so, Phobos is approaching Mars and it may crash into the planet in about 100,000,000 years. If Phobos has been in place ever since Mars was formed, about 4.6 billion years ago, then we are seeing the small satellite in the last 2 per cent of its lifetime.

Unfortunately, tidal drag would place the Martian satellites in their present position even if they were captured bodies that had originally formed in some other cloud of coalescing material.

If the satellites were captured as they passed by on their way from somewhere else to somewhere else, we could argue that the chances would be virtually certain they would then surely have elongated orbits that were highly inclined to the equatorial plane—at least, at the time of capture and for a considerable period afterward.

This is true, for instance, of the nine outermost satellites of Jupiter; of Phoebe, the outermost satellite of Saturn; of Nereid, the outermost satellite of Neptune—all of which astronomers confidently believe to represent captured bodies.

In earlier chapters I thought that the fact that Phobos and Deimos had circular orbits in Mars's equatorial plane was sufficient proof that they were not captured but were natural satellites, but I had not taken tidal drag into consideration. The captured satellites of the outer planets are so far away from the planets they circle and may have been captured so recently (astronomically speaking) that tidal drag has not yet had time to circularize and equatorialize them.

Similarly, whereas I have often thought that the fact that the Moon's orbit was not in Earth's equatorial plane was in itself ample evidence that it was captured, it may simply be that it happened to form in an odd orbit naturally out of the same cloud but is sufficiently massive compared to Earth so that tidal drag has not yet had time to complete the circularization and equatorialization.

The Martian satellites are, however, so small and so close to Mars and may have been captured so long ago, comparatively, that tidal drag *has* had a chance to produce circular orbits in the equatorial plane for them.

It follows that, from their orbits alone, it is impossible

to tell whether Phobos and Deimos formed in the neighborhood of Mars or were captured.

In that case, what about chemical composition?

There we may have something. Mars is a comparatively bright planet. Even discounting its atmosphere, it reflects quite a bit of light from its ruddy, oxidized, iron-rich soil.

The two satellites, however, are not like that at all. They are so surprisingly and unexpectedly dark (as I explained in Chapter 9) that their sizes are larger than had been calculated assuming a reasonably high albedo (or reflecting power).

This business of dark and bright, dark for the satellites and bright for Mars, seems to weight the chances in favor of the capture theory for Phobos and Deimos.

Captured from where?

The answer to that is easy. Beyond Mars is the asteroid belt, which is something like a primitive hangover from the infant days of the solar system. The asteroid belt, it would seem, is a place where the matter of the original cloud began to conglomerate but then remained as tens of thousands of separate bodies and never completed the process of forming a single planet.

This case of arrested development may have been the result of giant Jupiter's gravitational influence. It may be that Jupiter drew so much material out of the asteroid belt as to leave insufficient material to produce a total gravitational effect capable of forming a planet.

We can play with the notion, then, that two of the innumerable asteroids were captured by Mars, that they had elongated and inclined orbits to begin with, but that tidal drag put them into their present circular orbits in the equatorial plane.

Is that very likely? Would there be any asteroids likely to have orbits that would bring them close enough to Mars to be captured?

Absolutely! We know of a number of asteroids with

orbits that are elongated enough to be beyond that of Mars at aphelion, but closer than Mars (and Earth and Venus, too, in some cases, and in one extreme case, even Mercury) at perihelion.*

It is not at all beyond the bounds of possibility, then, that two asteroids had orbits permitting a very occasional rendezvous with Mars that made capture a possibility.

Next question. Does the fact that the satellites are dark fit the suggestion that they are of asteroidal origin?

The answer to that, surprisingly, is yes—or, at least, yes, if we go by the thinking of the last few years.

Consider the albedo of the Moon, for instance, which is 0.06. That means that it reflects 6 per cent of the light that falls upon it. This is what would be expected of a world without air and water, which presents to the Sun the kind of rock of which the crust of a normal world (that is, one like ours) would be built. Thus, the albedo of Mercury, which is likewise airless, waterless, and surface-rocky, is 0.07.

If we find airless worlds with albedos higher than that of the Moon or Mercury, we can assume they are not composed of normal rock. For instance, the four large satellites of Jupiter all seem to have albedos higher than that of the Moon or of Mercury even though it seems quite certain they possess no more than atmospheric traces. In the case of Io, the albedo is even as high as 0.54. We can assume that the surface of these satellites is composed of something shinier than rock—of ices of various sorts, of caked salts, and so on.

As to the asteroids—

Well, we'll go into the matter of asteroid albedos in the next chapter and educe from that still another reason why a human landing on the Martian satellites might, conceivably, be even more interesting than a landing on Mars itself.

* See "Updating the Asteroids," in my Doubleday collection *Of Matters Great and Small.*

11. The Real Finds Waiting

When I was a little boy, my parents, newly arrived in this country, did not know English and were helpless as far as guiding me in my reading was concerned. My father knew instinctively, however, that the material he himself sold for popular consumption—magazines of various sorts—would rot my mind if I allowed any of them to pass through my eyes (I disagreed with him there, but disagreeing with my father never got me anywhere until I reached college age).

Therefore, he got me a library card and told me to go there and get books. Since there was no one to tell me which books represented the classics, I didn't know which books to avoid.

I came upon Homer's *Iliad* at an early age, and not knowing I was *supposed* to read it, I read it. I couldn't pronounce any of the Greek names easily, but I could tell I had an exciting story there (especially since I had already worked my way through books on the Greek myths).

I particularly liked the translation by William Cullen Bryant, which I now realize must have been dreadful because he used the Roman rather than the Greek names of the gods. That didn't bother me then, however, and I picked it out of the library again and again and read it over and over. I sometimes returned to the first verse as soon as I had completed the last verse.

I had it almost memorized. At least, at the height of my infatuation, you might have stood behind me, opened the book to any page and read a single line, and I would have told you where that line came from, what preceded it and what succeeded it.

Alas, that was many years ago and I can't do that anymore, but I still know the *Iliad* in considerable detail and I know, for instance, that in the twenty-third book, which deals with the funeral games in honor of Patroclos, the prize for the shotput contest was the shot itself—a lump of iron.

Here is the passage, in William H. Rouse's translation: "Again Achilles brought out a lump of roughcast iron which that mighty man Eëtion used to hurl . . . 'Rise, you who wish to contend for this prize. Any man will have enough here to use for five revolving years . . . No shepherd or plowman will need to visit the city for iron, there will be plenty at home.' "

Iron? Of what value is iron? And how could a small lump (for even one that was large enough for only a strong man to throw had to be a small lump) last for five years?

It took me a while to solve that mystery. The Greeks at the time of the Trojan War (c. 1200 B.C.) were living in the Bronze Age, for the metallurgical techniques required to produce copper and tin, the components of bronze, from their ores were comparatively simple. The metallurgical techniques for producing iron from its ores were more difficult and not developed until the centuries immediately following the Trojan War. It wasn't till 800 B.C. that iron became sufficiently common for routine use in military armor and weapons.

In that case, where did Achilles' iron come from? Was it a bit of anachronism on Homer's part, assuming Homer sang his epics several centuries after the Trojan War. No. For one thing, Homer was surprisingly unanachronistic and, for another, iron was indeed available and had actually been used in small quantities since at least 3000 B.C.

How so? Because there was some iron that didn't have to be smelted out of its ore, since it was to be found in the ground as such. In fact, it was better than mere iron. It was mixed with nickel and was the equivalent of alloy steel. It was strong and tough and, used very abstemiously, could make points and edges for tools and weapons made mostly of other materials.

That iron was meteoric in origin, and occasional findings must have been cause for jubilation. There are only some 550 such iron "meteorites" that have been found altogether by modern man, these findings coming from all over the world except in areas where early civilizations used metals. There the finds have long ago been gathered up and used.

The largest find presently known is still in the ground in Namibia, in Southwest Africa. It is estimated to weigh about 66 tons. The largest known iron meteorite on display is at the Hayden Planetarium in New York and weighs about 34 tons.

Meteorites fall from the sky and were originally "meteoroids" in independent orbit about the Sun. (It took scientists a long time to believe that matter could fall from the sky, but that is another story for another day.)

Occasionally, a meteorite was actually seen to fall from the sky and to land on Earth by fortunately placed individuals. It could then seem to be a holy object sent by the gods. The "black stone" in the Kaaba, holy to Muslims, may be such a meteorite. The original object of veneration in the temple to Artemis at Ephesus may have been another.

Actually, meteorites are not iron in every case. There are also "stony meteorites," made up of silicate materials rather like that of the crust of the Earth.

For a long time it was considered that stony meteorites were rather unusual and that it was the iron meteorite that was typical. Of the stony meteorites that were discovered, a disproportionately large number were discovered, in the 1930s, in Kansas.

I can well remember John Campbell, the late editor of *Astounding Science Fiction*, telling me of this very unusual fact and saying that scientists couldn't explain it. I even seem to remember he wrote an editorial on the subject and that a novella was written called *The Gods Hate Kansas* which I think dealt with the matter. It didn't appear in

Astounding, though, but in the November 1941 issue of *Startling Stories*.

I could not explain the mystery to John Campbell at the time, because I was not smart enough, but at least I was smart enough to maintain that I didn't think it was a mystery and that the explanation, when it came, would prove simple.

And it did. There just isn't any pure iron occurring naturally in the Earth's crust, you see; thanks to the high-oxygen atmosphere, all crustal iron is in the form of oxides and silicates. Any iron you find on Earth's surface now is either man-made or meteoric; and in the days before the Iron Age, any iron you found was meteoric. Iron meteorites were therefore easy to recognize if found, for if an object was iron, it was automatically a meteorite.

Not so stony meteorites. Earth's land surface is richer in stones, generally speaking, than in anything else. Whether a stone has fallen from the sky or has been there all along as part of the planet cannot be told at first glance, and no casual human being would even stop to wonder at one more rock. (An expert might suspect, of course, by various outward signs, and confirm the suspicion by a close study of the stone's structure, for even a stony meteorite is likely to have inclusions of metallic iron.)

On the other hand, in an area such as Kansas, where the soil is deep and where surface rocks are virtually non-existent, the presence of any rock is unusual enough to call attention to itself, and stony meteorites are much more easily located there than they would be, let us say, in Vermont.

The gods don't hate Kansas; they hate all places equally. It's just more noticeable in Kansas. Nevertheless, I'm willing to bet that the gods-hate-Kansas idiocy is still peddled and believed by mystery-loving ignoramuses.

But what if a meteoroid is seen to fall and an object is then identified as that just-fallen meteorite? In the majority of cases, it turns out to be a *stony* meteorite. In other words, the association of meteorites with iron and the assumption that iron is typical of meteorites are just the results of the accident that an iron meteorite is much easier to identify as such than a stony meteorite is.

In actual fact, it may be that 90 per cent or more of the meteorites on Earth are stony (or stony-iron, a kind of intermediate classification) and less than 10 per cent are iron. This makes sense, actually, since although iron is the commonest of the relatively massive atoms, it is not as common as the lighter atoms making up the silicates of stone.

Where do the meteorites come from? The consensus seems to be that they are from the asteroid belt. The original cloud of matter out of which the planets of the solar system formed never managed to form a single planet in the space between the orbits of Mars and Jupiter, perhaps because of the gravitational effect of giant Jupiter (see Chapter 10), but it may have formed a collection of objects that included several dozen fairly large objects of subplanetary size.

The largest may have been large enough to melt on formation and to differentiate into objects with iron-rich cores and stony mantles. In the early days of the solar system there may have been collisions of these with smaller pieces so that there would be crater formation. There might also have been shattering and, in view of the low gravitational intensities involved, small asteroidal pieces may have flown off in every direction.

Some of the larger pieces may have, through the accident of random hits by randomly sized objects, disintegrated altogether, so that even bits of the iron core might, in that case, have flown off in every direction, too.

The final result would be the asteroid belt, with the orbits of its members having evolved and stabilized over the billions of years into mutual non-interference. Some of the asteroids, however, may have achieved orbits outside the general belt in either direction. A number move within the orbit of Mars and approach the Sun at perihelion more closely than Mars does—or even than Earth and Venus and, in one known case, Mercury do.

Those intra-Martian asteroids which we spot are the fairly large ones, a kilometer or more across, but there must be numerous such asteroids that are smaller fragments, only meters or even centimeters across, and these are the meteoroids. It is not likely that large asteroids will hit us because they are so few (though it is not impossible either), but meteoroids would hit us frequently since there are so many. (There are also dust-sized "micrometeoroids" in countless numbers, but they probably originate from disintegrating comets.)

If meteoroids come from the asteroid belt and if stony ones are more common than iron ones because the former are made up of lighter atoms, ought there not to be some that are made up of the lightest and most common atoms of all, such as hydrogen, oxygen, carbon, nitrogen, sulfur, and so on? Admittedly, these very light atoms, in themselves or in simple compounds, might be difficult for objects with small gravitational fields to hold onto, but at the low temperatures prevailing in the asteroid belt, especially in the farther reaches away from the Sun and nearer Jupiter, the difficulty should not be an insuperable one.

The answer to that takes us back to meteorites.

Over 90 per cent of the stony meteorites contain "chondrules," small round bodies consisting of magnesium-iron silicates, the term coming from a Greek word meaning "grain." Meteorites possessing them are called "chondrites."

Every once in a while, a chondrite is found which contains a significant amount of carbon. Not much, only about 2 to 4 per cent, but that is a lot for a meteorite, and such objects are called "carbonaceous chondrites."

Only about twenty carbonaceous chondrites have been seen to fall and have then been picked up. One important fall came on September 28, 1969, when an object exploded over the town of Murchison, Australia, and showered fragments over an area of 36 square kilometers (14 square miles). Eventually, 82.7 kilograms (182 pounds) were picked up. In 1950 there had been a smaller fall near Murray, Kentucky.

Such carbonaceous chondrites made scientific news because the carbon existed within them in the form of compounds such as fatlike hydrocarbons and, more interesting still, as amino acids, which are the building blocks of proteins. (Do not, however, jump to conclusions. There are unmistakable chemical signs that such organic compounds were made by processes that did not involve life—at least, not life as we know it.)

The carbonaceous chondrites are divided into three classes—I, II, and III, of which class I is richest in the light elements and compounds, in carbon, in water, in sulfur, and so on. Except for the very lightest elements, the carbonaceous chondrites seem to have an elementary composition similar to that of the Sun. They might be the nearest objects, chemically, between Jupiter and the Sun that strongly resemble the original cloud of dust and gas out of which the solar system was formed.

It would seem, then, that the larger asteroids, if they were differentiated, would not only have an iron core and a stony outside, but a light-element crust. Therefore, just as there are more stones than irons, there might be still more carbonaceous objects among the meteoroids.

Why, then, do so few carbonaceous chondrites strike the Earth?

Perhaps many do, but whereas the iron and stone ones are tough and have high melting points, the carbonaceous chondrites are objects that easily crumble and pulverize. They are simply too fragile to survive the hot passage through the atmosphere unless they are unusually large.

This is a shame, since the carbonaceous chondrites, because of the complex compounds they contain and their apparent closer approach to the composition of the original nebula, could have a great deal to tell us about the early history of the solar system and of life.

It may be, though, that carbonaceous chondrites are

intrinsically rare and that that's just tough on us. Well, let's see.

The asteroids generally are too small and too far away to be seen as disks with even good telescopes. To estimate their size, therefore, the best that can be done is to suppose they are made of rock and have the albedo (or reflecting power) of stony bodies. Dark rock, like that of the Moon, reflects only 6 per cent of the light that falls upon it, but lighter rock could easily reflect up to 13 or even 16 per cent of the light. If we assume some particular albedo for a particular asteroid, then, knowing its distance and brightness, we can calculate its diameter.

In the 1890s, however, the American astronomer Edward Emerson Barnard, whose eyesight was legendary, using the 40-inch Yerkes refractor, in William's Bay, Wisconsin (then, and today, the largest and best of its kind in the world), thought he could just make out the disklike shapes of the first four asteroids to be discovered—these having been discovered because they were brightest and therefore, presumably, the largest.

From the disk of Ceres, he calculated its diameter to be about 750 kilometers (456 miles), for instance. This seemed to fit in well with the albedo expected of a rather light stone. Vesta, however, seemed to have a considerably higher albedo, considering the size Barnard made out for it, and that seemed to pose a mystery.*

However, even Barnard's eyes could just barely make out the disks of those four asteroids, and the measurements of the diameters were therefore pretty chancy.

What if, instead, the problem were tackled from the other end and the albedo were determined by new and sophisticated techniques, such as by studying the polarization of the reflected light when the asteroid was in different positions in the sky so that the Sun/asteroid/Earth angle changes? Or perhaps if the amount of sunlight the asteroid

* See "Little Lost Satellite," in the Doubleday collection *The Solar System and Back*.

absorbed could be determined by measuring its radiation in the infrared?

By the mid-1970s it began to appear that many of the asteroids were darker than had been thought and had albedos considerably lower than that of even the comparatively dark-rocked Moon. These asteroids reflected less light than had been thought and therefore had to be larger than had been estimated.

Ceres, for instance, reflected only 5.4 per cent of the light that fell upon it and its diameter had to be raised from Barnard's earlier estimate to a new diameter of 1003 kilometers (623 miles). This makes it an even more respectable world than the one I described in "Little Lost Satellite" (see footnote on p. 144).

Ceres' diameter is nearly 30 per cent that of our Moon. Its surface area would be 3,140,000 square kilometers (1,190,000 square miles), or about 30 per cent that of our fifty states.

A number of asteroids have albedos even lower than that of Ceres (some as low as 2 per cent). One way in which such darkness might be explained is to suppose that the asteroidal surface was of basalt. Basalt is a dark igneous rock, however, formed under considerable heat and pressure, and on Earth it is formed chiefly through volcanic action. But it is hard to imagine very much volcanic action on even the larger asteroids, and no one seriously advances basalt as the explanation.

The alternative is the dark element carbon. In short, many of the asteroids seem to be akin to the carbonaceous chondrites that occasionally land on the Earth, although, to be sure, it may be only their crusts that are carbonaceous. For that reason, the dark asteroids may be classified as C.

Brighter asteroids, with albedos in the region of 0.09, may be metallic (M); still brighter ones, 0.13 to 0.16, may be stony (S); and some, like Vesta, 0.23, don't fall into the neat pigeonholes and they are usually listed as "unclassified" (U).

From albedo data, Table 8 lists what are now thought to be the twelve largest known asteroids.

The first thing that springs to the eye in Table 8 is that of the dozen largest asteroids eight, or 67 per cent, are in class C.

On the whole, it is expected that the farther from the Sun one goes in the asteroid belt, the cooler the temperatures at the time of asteroid formation and the greater the chance for the light elements to accumulate. Furthermore, the best determinations are made on the brighter asteroids, which means that for two asteroids of equal size, the one that is nearer the Sun or the one that is of higher albedo (or both, of course) is more likely to be measured accurately. Therefore, a table like Table 8 automatically favors asteroids that are not C, so that the actual percentage of C asteroids is even higher than would seem from the table.

In fact, the best surmise from the available data is that some 88 per cent of all the asteroids are in the C class. Of the asteroids in the inner reaches of the asteroid belt, where initial temperatures were highest, perhaps only 50

TABLE 8—THE TWELVE LARGEST ASTEROIDS

Order of Discovery	Name	Diameter		Classi- fication
		Kilometers	Miles	
1	Ceres	1003	623	C
2	Pallas	608	378	U
4	Vesta	538	334	U
10	Hygeia	450	280	C
31	Euphrosyne	370	230	C
704	Interamnia	350	218	C
511	Davida	323	201	C
65	Cybele	309	192	C
52	Europa	289	180	C
451	Patientia	276	172	C
15	Eunomia	272	169	S
16	Psyche	250	155	M

per cent are Cs, but in the outer reaches the percentage of Cs may rise to 95 per cent.

Now, then, granted that the carbonaceous layer is only on the outer surface of the asteroid and may even be thin, it is the outer surface that is precisely the portion that would most easily be sent flying in shattering and fragmenting collisions.

Therefore, there is nothing in the least strange in arguing optimistically that a large proportion, perhaps even more than 50 per cent, of the meteoroids in the inner solar system are carbonaceous chondrites and that the same proportion of those striking our atmosphere are. It is only the difficulty of surviving atmospheric passage that is giving us trouble.

Well, then, granted we enter the Space Age in full, we have several choices of how to locate carbonaceous chondrite material if we wish to use it to study the origins of the solar system and of the possible chemical routes to life.

The first and most obvious alternative is to engage in "space prospecting" in the vicinity of Earth. In other words, we might keep our eyes open for small objects and try to catch them and study them. Our situation would then be like that of the Greeks of Achilles' time, keeping their eyes open for any meteorites that might be lying around, or like that of the nineteenth-century gold prospectors in California and the Yukon, panning streams for fragments of gold.

It would be exciting, but the percentage of failures would be very high.

A second alternative would be to chase after known asteroidal bodies in Earth's vicinity for at least part of their orbits—Eros, Icarus, Geographos, and Toro.

Unfortunately, all four of these have albedos high enough to indicate they are stony in nature.

In fact, we might pessimistically argue that any meteoroid with an orbit that penetrates the inner solar system is liable to lose, bit by bit, any carbonaceous outer layer it might have had, through evaporation in the solar heat.

It might be that, after all, the pickings are few out here near Earth and that the reasons for the comparative rarity of the carbonaceous chondrites that reach our surface is not only their fragility to atmospheric passage, but also their relative rarity in nearby outer space as well.

In that case we might have to pass on to the third alternative and carry on our search in the asteroid belt.

If we could reach Ceres, we would have a sizable body, the sixth largest solid body in the solar system inside the orbit of Jupiter, with a surface that is very likely carbonaceous, and it might tell us quite a bit.

Ceres is far away, however. Its average close approach to us is about 265,000,000 kilometers (164,000,000 miles), or some four times the average close approach of Mars.

We might do a little better by choosing a closer, if smaller, asteroid. The dark asteroid Feronia is about 75,000,000 kilometers (46,000,000 miles) closer than Ceres is, on the average—an improvement but not enough of one.

Anything else? Yes, there remains a possible fourth alternative.

In the previous two chapters, I discussed Phobos and Deimos, the satellites of Mars, and I mentioned that they were considerably darker and considerably lower in density than had been expected. They are therefore thought likely to be captured asteroids, and if so, they are surely of type C and are carbonaceous chondrite in character.

There's no question that Phobos and Deimos can be reached. We've reached Mars and, astronomically speaking, Phobos and Deimos are in the same apartment house.

Why then concentrate entirely on Mars? Grant the enormous interest we must have in that planet, but let us not forget the satellites.

Those satellites are the nearest examples of material left totally unchanged since the end of the crater-forming period 4 billion years ago. Consider, too, that they may be possessed of small quantities of water and organic material which, for 4 billion years, may well have been undergoing chemical evolution without having been disturbed by the heat of passing through Earth's atmosphere or the impact of having collided with Earth's solid mass.

This is not to say that there is the faintest possibility of life-forms on either satellite, but the non-living organic material may give us some hints as to how and what took place in the way of chemical changes (admittedly, in the absence of an atmosphere and ocean), and who knows how useful that might be?

It would be the height of irony if, while the Viking landers were busy testing Mars for living organisms in its soil and finding no organic matter there, the real finds were waiting to be made in the tiny asteroids wheeling overhead.

12. Rings and Things

I imagine that some of you must by now have noticed that there is a science fiction magazine on the stands by the name of *Isaac Asimov's Science Fiction Magazine* (*IASFM*). Those of you who have seen it may conceivably be wondering from month to month whether my backlog of essays will give out and whether I will then cease to appear in the hallowed pages of *F & SF*.

The answer to that is, No chance!

I am a permanent fixture at *F & SF* until I pass on to the great typewriter in the sky or until Ed Ferman gives me the boot, whichever comes first. Since this is the twentieth essay published in *F & SF* since *IASFM* first appeared, you can see that I mean it.

Meanwhile, though, more than one person has written to ask if this double role of mine does not represent a conflict of interest.

The answer to that is, Certainly not!

When Joel Davis of Davis Publications suggested the new magazine to me in early 1976 and told me that it had to have my name in the title to match the names in the titles of his other fiction magazines, I told him that I could on no account agree to this if it meant abandoning my *F & SF* column. Joel promptly agreed to allow the column to continue.

I then consulted Ed Ferman of *F & SF* and Ben Bova of *Analog* for I had to find out whether either editor would have serious objections to the new venture. Both are close friends of mine of many years standing and you may take it for granted that I would do nothing to hurt either one.

Both Ed and Ben assured me that a new, strong magazine would provide another short-story market and would in that way encourage writers, especially new young writers. This would strengthen the field generally and work to the benefit of the older magazines as well.

So you see, everything I did was open and aboveboard and I made sure of obtaining the consent and blessings of all concerned before making a move.

Having straightened out that matter, let's forget it and go on about our business. We will take up the outer solar system where exciting things have been happening—

I once discussed Saturn's rings in my essay "Little Found Satellite" and said: "There was simply nothing like Saturn's rings in the heavens, *and there still isn't*. It is absolutely unique among all the heavenly features we can see."*

That was ten years ago but even then I might have asked why Saturn's ring system should have been unique. Let's think about it now.

In 1849 the French astronomer Édouard Roche showed that any sizable satellite with the same density as the planet it circled would be broken up by gravitational influences if it were closer to the planet than 2.44 times the planetary radius. This is called the "Roche limit."† Any satellite inside this limit would be broken up into irregular chunks of comparatively small size, and these, as a result of tidal effects and mutual collisions, would eventually take up more or less circular orbits in the equatorial plane of the planet.

* In my Doubleday collection *The Solar System and Back*.
† Roche assumed the satellite would be held together by gravitational effects only. Since satellites are usually held together by electromagnetic bonding too, they might get a little closer than the Roche limit before breaking up.

Conversely, if a vast array of fragments were circling a planet at a distance less than the Roche limit, they would be unable to coalesce into a sizable satellite.

The equatorial radius of Saturn is 60,400 kilometers (37,500 miles), so the Roche limit for Saturn is 147,400 kilometers (91,500 miles).

Saturn's ring system extends from an inner edge at 72,000 kilometers (45,000 miles) from the center of Saturn to an outer edge at 137,000 kilometers (85,000 miles) from the center of Saturn; or from 1.2 to 2.3 times Saturn's radius.

Janus, the closest known sizable satellite of the Saturnian system, is at a distance of 159,500 kilometers (99,000 miles) from Saturn's center, or 2.6 times Saturn's radius.

In other words, the entire Saturnian ring system lies within the Roche limit, while Janus lies just beyond the Roche limit.

Is this a situation that should exist for all planets? When a planet forms might there not be an extension of material, gradually thinning, for millions of kilometers about itself? Might not the material outside the Roche limit form a series of satellites, and the material within the Roche limit form rings—as apparently was the case with Saturn?

In the case of the inner planets, this proposed scenario might fail, since the solar wind could be strong enough, at near distances to the Sun, to sweep away the outer, less dense portions of the cloud that was coalescing to form a planet, leaving nothing out of which satellites might form.

Thus, Mercury and Venus have no satellites. Earth has the Moon, but the latter is of dubious origin and may well have been an independent planet captured by the Earth well after the origin of both. As for Mars, it has two satellites, but both are tiny and are almost certainly captured asteroids (see Chapter 10).

That leaves us, though, with the planets of the outer solar system. They are so far from the Sun that the solar wind, enfeebled by distance, would have done little to sweep away matter in the outskirts of the forming planet.

Then, too, the outer planets, at their distance from the Sun, would be frigid enough to collect the hydrogen, helium, and hydrogen compounds that the inner planets in the Sun's neighborhood would be too warm to retain. The outer planets would therefore grow to giant size and would develop gravitational fields that could more effectively hold on to the uncoalesced matter in its neighborhood.

The outer planets do, therefore, have satellites. And why not rings as well?

We might argue that it makes sense to suppose that the circumplanetary cloud would extend all the way down to the planet itself and be denser and denser as such a cloud approaches the planet. There would then certainly be more matter per cubic kilometer within the Roche limit (or what would eventually become the Roche limits as the planet completes its coalescence) than outside it. Such an argument would favor rings for every outer planet.

Suppose we try a counterargument. There would naturally be a tendency for the matter in the circumplanetary cloud to settle down onto the planet itself. The closer to the planet, the more likely that would be, so it might be that a planet, in forming, sweeps out those regions within the Roche limit. In that case, satellites form, but no rings.

Consider Jupiter. It has thirteen known satellites, possibly fourteen. Of these, all but the innermost five are captured asteroids. Of the five that were probably formed out of the same cloud that formed Jupiter, four are satellite giants, of the order of size of our own Moon. All of these are well outside the Roche limit. For Jupiter, the Roche limit is 174,460 kilometers (108,400 miles) and the innermost of the giant satellites, Io, is at a distance 2.4 times that of the Roche limit.

The fifth satellite, Amalthea, is, however, closer to Jupiter than Io is, and circles the planet at a distance of 180,500 kilometers (112,000 miles). This is only 1.035 times the Roche limit. If Amalthea were only 6000 kilometers (3600 miles) closer to Jupiter, tidal effects would break it up and it would form a set of rings about Jupiter. So close is Jupiter to being a ringed planet.

However, Amalthea is no giant satellite. It is perhaps 150 kilometers (about 90 miles) in diameter, only 1/10,000 of the mass of Europa, the smallest of Jupiter's giant satellites. Apparently, so much of the matter in the Amalthean vicinity was picked up by Jupiter in the days of the planet's formation that what was left over sufficed only for a moonlet.‡

But then, since Jupiter has no rings, why should Saturn have them? And if Saturn has rings, why should Jupiter have none?*

We might argue that Saturn just happens to be in the "just right" position.

When the solar system formed, the density of the material in the original cloud must have decreased steadily with distance from the Sun. That's why Jupiter is the largest of the four outer giants, 3 1/3 times as massive as Saturn. Saturn is, in its turn, about 6 times as massive as either Uranus or Neptune.

It might be argued, therefore, that Jupiter is so massive and its gravitational field so intense that it swept out the matter within its Roche limit very effectively. Uranus and Neptune, on the other hand, had so little matter about themselves that there wasn't enough to form noticeable rings after the planets' comparatively feeble gravities were through sweeping up what they could.

Saturn, however, was just close enough to the Sun to be surrounded by considerable matter and just far enough away to be too small to sweep up its near-vicinity effec-

---

‡ Amalthea's mass is only about 1/5600 that of Saturn's rings. Amalthea wouldn't make much of a ring system even if it did break up.

* Since this essay was written, the Jupiter probe Voyager 1 has detected thin and inconspicuous rings around Jupiter.

tively. It therefore left sufficient matter behind within the Roche limit to form its splendid rings.

But wait! I said Uranus and Neptune might not form "noticeable" rings. After all, it isn't a matter of gorgeous rings, like Saturn's, or none at all. Might there not be small, unpretentious ones? Uranus and Neptune are far distant and hard to observe. How certain can we be, then, there are no rings at all out there? Narrow ones? Dim ones?

I don't know that the question ever arose, but in 1973 a British astronomer, Gordon Taylor, calculated that Uranus would move in front of a ninth-magnitude star, SAO 158687, in the constellation Libra, on March 10, 1977.

On that day, James L. Elliot and associates from Cornell University observed the occultation from an airplane that took them high enough to minimize the distorting and obscuring effects of the lower atmosphere.

The notion was to observe just how the starlight was affected as Uranus reached and began to move in front of the star. The starlight would penetrate Uranus's upper atmosphere and would in this way yield information about its atmospheric temperature, pressure, and composition.

But some time before Uranus reached the star, the starlight suddenly dimmed for about seven seconds and brightened. Then, as Uranus approached still more closely, there were four more brief episodes of dimming, for a second each. Uranus eventually passed in front of the star, and as the planet moved away on the other side, there was the same dimming of starlight in reverse—four times for a second each and then a fifth time for seven seconds. (Other astronomers studying the occultation also observed the dimming effect.)

Something was obscuring the star, something in Uranus' vicinity. At first, Elliot thought it was a satellite or several of them, but after he had a chance to study the data and notice the symmetrical nature of the dimming, he knew it had to be rings. Uranus had to have a ring system consisting of at least five rings one inside the other.

Why did it take so long to discover the rings of Uranus?

First, Uranus is distant. The total distance light must travel from the Sun to a planet to Earth is four times as great for Uranus as for Saturn, so that, all things being equal, the Uranian ring system would be only 1/16 as bright as the Saturnian system.

Second, all things are not equal. The Uranian rings are very narrow. The thin ones that obscured the star for only a second apiece are each about 12 kilometers (7.5 miles) wide, and the wide one might be 85 kilometers (50 miles) wide. The total width of the Uranian rings is 170 kilometers (105 miles) as compared with a total width, allowing for gaps, of 64,000 kilometers (40,000 miles) for the Saturnian rings.

Third, the material in the Uranian rings is more thinly strewn than that within the Saturnian rings. The star, in passing behind the Uranian rings, was not blanked out altogether, but was merely dimmed.

Fourth, the Uranian rings are not of the same composition as the Saturnian rings. The Saturnian ring particles are highly reflective, reflecting more than half the light that falls on them, so that we can say the rings are almost surely composed of icy particles. The Uranian ring particles, on the other hand, are dark and reflect less than a twentieth of the light that falls upon them. They must be made of rocky material, and dark rock at that.

Lumping all these differences—distance, width, density, and reflectivity—we might estimate that the total light reaching us from the Uranian rings may be only 1/3,000,000 that which reaches us from the Saturnian rings.

No wonder, then, that it took us so long to spot the Uranian rings. But for the lucky accident of the occultation, we might still have no idea of their existence.

And what about Neptune now? If Neptune has rings, they might be an even shabbier set than Uranus'. And, as a matter of fact, Neptune occulted a star in 1968. The occultation was observed and no dimmings of starlight in Neptune's neighborhood were noted.

Still, our solar system now has three ringed planets and it is likely that the general phenomenon of ringed planets in the Universe is much more common a thing than had been thought possible only two years ago.

The finding of the Uranian rings was only one of the two exciting discoveries of 1977 in the Saturn-Uranus section of the solar system, however, so let's carry on.

In my essay "Updating the Asteroids" (in my Doubleday collection *Of Matters Great and Small*) I concentrated on those asteroids whose orbits carried them out of the asteroid belt and inside the orbit of Mars.

What about asteroids which are unusual in the other respect and which move out of the asteroid belt beyond the orbit of Jupiter. Let's call those "hyperasteroids," a term I have just invented.

The first borderline case of a hyperasteroid was discovered on February 22, 1906, by the German astronomer Max Wolf. It was asteroid #588 and Wolf named it Achilles. It was the first of the Trojan asteroids to be discovered. This is a group of asteroids that travel in Jupiter's orbit and move in step with the giant planet. Some are in a stable position (L4) which is 60° ahead of the planet, and some are in a stable position (L5) which is 60° behind the planet (see "The Trojan Hearse," in my Doubleday collection *Asimov on Astronomy*).

About fifteen Trojans have been spotted in the L4 position and five more in the L5 position, but it has been estimated that there are seven hundred or more Trojans altogether. Naturally, we see only the largest—those that are 100 kilometers (60 miles) or more in diameter.

The Trojan asteroids do not remain exactly in L4 or L5. Their original motion and the gravitational influence of Saturn cause them to oscillate about L4 or L5 in a complicated way as they move about the Sun. Such oscillations can be quite large, and an unlucky concatenation of perturbations may move a Trojan so far from L4 or L5 that it cannot regain its post and it then ceases to be a Trojan. On the other hand, non-Trojan asteroids moving nearby may, under favorable conditions of perturbation, be captured and become Trojans. On the whole, then, the Trojan positions probably lose a few, win a few, and remain about as they are.

A Trojan asteroid, if ejected from its position, is quite likely to have part of its new orbit well beyond that of Jupiter. It may then spend most of its time beyond Jupiter since any body moves more slowly as its distance from the Sun increases.

In 1920 the German astronomer Walter Baade discovered asteroid #944, which he called Hidalgo and which may well be an ejected Trojan.

Hidalgo has a high orbital eccentricity of 0.66. At perihelion, the point in its orbit that is closest to the Sun, it is only about 300,000,000 kilometers (190,000,000 miles) from the Sun and is then neatly within the asteroid belt. At aphelion, however, when it is farthest from the Sun, it is 1,450,000,000 kilometers (895,000,000 miles) away, or just about the distance of Saturn from the Sun.†

Hidalgo is a clear case of a hyperasteroid, with an orbital period of 13.7 years, the only inhabitant of the asteroid belt to have an orbital period that is longer than that of Jupiter.

There is another group of asteroids that accompany Jupiter even more closely than the Trojans do—that are neither before nor behind it, but are right there with it in the form of captured satellites.

Eight of these are now known. They were discovered more or less in order of decreasing size and, therefore, brightness.

The first to be discovered is usually known as J-VI,

---

† Hidalgo's orbital inclination is 43°, however, so that if we count in the third dimension, it never approaches any closer to Saturn than Earth does.

because it was the sixth to be seen. It was discovered in 1904 by the American astronomer, Charles Dillon Perrine.

Of Jupiter's captured satellites, J-VI is the closest to Jupiter. Its mean distance from Jupiter is 11,470,000 kilometers (7,100,000 miles) and it has an orbital period of 0.69 years. Its diameter is about 120 kilometers (75 miles).

The latest to be discovered is J-XIII. It was detected in 1974 by the American astronomer Charles Kowal and is only about 8 kilometers (5 miles) in diameter.

The known satellite farthest from Jupiter is J-IX, which was discovered by the American astronomer Seth Barnes Nicholson in 1914. It has a diameter of about 15 kilometers (9 miles). Its mean distance from Jupiter is 23,700,000 kilometers (14,600,000 miles) and its orbital period is 2.07 years. No other satellite in the solar system is so far from its primary or has so long an orbital period.

Undoubtedly, there are other captured asteroids in the farther reaches of Jupiter's far-stretching influence, but even if there are, they are not true hyperasteroids. Their average orbit is precisely that of Jupiter, as is also true of the Trojans.

Even Hidalgo, which spends most of its time beyond Jupiter, returns each orbit to the haven of the asteroid belt.

The question is, Are there any asteroids that are purely hyperasteroidal, with orbits lying *entirely* beyond that of Jupiter?

It seems to me there must be. In fact, we know of two cases.

In 1898 only eight satellites were known to be circling Saturn. In that year, the American astronomer William Henry Pickering discovered a ninth on a photographic plate. It was the first satellite to be discovered by photography. Pickering named it Phoebe, after a Titaness in the Greek myths.

The mean distance of Phoebe from Saturn turned out to be 12,900,000 kilometers (8,030,000 miles), or 3.6

times the distance of the next most distant satellite, Iapetus. Phoebe's orbit was more eccentric and considerably more inclined to Saturn's equatorial plane than was true for any other Saturnian satellite. The general astronomical opinion then is that Phoebe is a captured asteroid.

As in the case of the captured Jovian satellites, Phoebe is small, with a diameter of only about 300 kilometers (180 miles). It is quite possible that there are other, smaller objects in the outer reaches of the Saturnian system which we don't see only because they aren't as large as Phoebe.

Where did Phoebe come from, though? Was it an ejected Trojan that blundered its way into Saturn's grasp? If you're content with that explanation, then let's pass on to another object.

Neptune has a large satellite, Triton, with a diameter of about 4000 kilometers (2500 miles). It was discovered within days of the discovery of Neptune itself, in 1846. Just over a century later, in 1949, the Dutch-American astronomer Gerard Kuiper, whom we met in Chapter 9, located a second Neptunian satellite, which he named Nereid.

Nereid, which is about as large as Phoebe, has a very unusual orbit. Its mean distance from Neptune is 5,560,000 kilometers (3,450,000 miles), or fifteen times as great as that of Triton, but that is not all. Nereid's orbit is an elongated ellipse and has an eccentricity of 0.75, far greater than that of any other satellite in the solar system.

When it is at its closest point to Neptune, Nereid is only 1,400,000 kilometers (900,000 miles) away, but at the other end of its orbit it recedes to a distance of 9,500,000 kilometers (6,000,000 miles).

Nereid bears all the earmarks of a captured asteroid, but, in that case, where was it captured from? It is simply stretching things beyond the point of belief to suppose that Nereid was ejected from the asteroid belt with enough force to send it out to Neptune's orbit.

It seems to make much more sense to suppose that hyperasteroids exist in sizable numbers throughout the outer solar system and that it is only their great distance from us that makes it so hard to spot them. Each of the giant planets may have far more captured asteroids as part of their satellite system than we can see from Earth, and each of them—and not Jupiter only—may have asteroids in their L4 and L5 positions.

Nor is it just that hyperasteroids are very hard to see. Astronomers haven't really been looking for them. In searching for asteroids, they look for objects moving at the characteristic speed of the asteroid belt, a speed faster than that of Jupiter. Hyperasteroids would be moving at much slower speeds and might well be overlooked for that reason.

On November 1, 1977, for instance, Kowal, the discoverer of J-XIII, was studying his photographic plates. He was searching for new Trojan asteroids, which, of course, move across the sky more slowly than any other asteroids except for Hidalgo in the farther part of its orbit.

He detected something that looked about the right brightness and was moving slowly—but it was moving *too* slowly. It was moving at only one third the speed expected of a Trojan asteroid and therefore it would have to be something much farther away.

It was moving so slowly, in fact, that it would have to be somewhere in the vicinity of the orbit of Uranus, and, allowing for that distance it would have to be something of asteroidal size, along the order of the size of Phoebe or Nereid.

It was not part of the Uranian satellite system, however. In fact, it was nowhere near Uranus which, at the time, was almost exactly on the other side of the Sun.

Kowal followed it for a period of days, worked out an approximate orbit, then started looking for it on earlier photographic plates covering the regions where it ought to have been. He located it here and there, and eventually enough positions were plotted for an accurate orbit to be worked out.

Kowal named the new object Chiron, after the best-known centaur in the Greek myths.

Chiron turned out to have quite an elliptical orbit, with an eccentricity of 0.38. At aphelion it is 2,800,000,000 kilometers (1,740,000,000 miles) from the Sun, which is about as far as Uranus gets. Chiron was at aphelion in November 1970 and has been moving closer to the Sun over since.

At perihelion, where Chiron was in August 1945 and where it will be again in February 1996, Chiron is 1,270,000,000 kilometers (790,000,000 miles) from the Sun and it will then be slightly closer to the Sun than Saturn is.

In short, Chiron seems to gallop in true centaur fashion between the orbits of Saturn and Uranus. Its orbital inclination is 6.9°, though, so that there is no danger of its colliding with either planet. It never gets closer than 150,000,000 kilometers (95,000,000 miles) to Saturn and misses Uranus by considerably more than that. Its orbital period is 50.7 years.

Chiron is the first pure and independent hyperasteroid to be discovered, but I feel it is only the first of a large group, and once we establish astronomical observatories in space (say, on Phobos or Deimos) and thoroughly computerize our search, we will find hyperasteroids by the thousands.

E STARS

13. Proxima

I made *The New Yorker* once.

Not I, actually, but a reviewer for some Alabama newspaper who accomplished the feat on my account. He was reviewing a book of mine and in the review said something to this effect—

"Isaac Asimov types 90 words a minute and it is to this that he attributes much of his profligacy."

Someone then sent that to *The New Yorker*, where it appeared with the comment: "So do we, but we are the soul of decorum, nevertheless."

I laughed when I saw it because it was an example of the errors made by those good souls who like to use a level of vocabulary to which they cannot truly aspire. What the reviewer meant to say was that it was to my typing speed that I attributed much of my "prolificacy" or "prolificity."

"Prolificacy" would refer to my prodigious output, the existence of which I admit and which has an obvious connection with speed-typing. "Profligacy" refers to my dissolute conduct, the existence of which I don't choose to admit and which, in any case, has no connection at all with speed-typing (except that the latter might leave me some spare time).

Within a week of its appearance, though, I received five copies of *The New Yorker* item, each carefully clipped out, from five different acquaintances—each of the five being young women of surpassing beauty. I'm not sure of the significance of that.

The incident does bring to mind the loose use of words

even by those whose vocabularies have ample scope for the purposes at hand.

To "discover" something is to bring to light an object that was already in existence but previously unknown. To "invent" something is to produce something that has never existed before. The American continents can be discovered but not invented.

Yet can we speak of "the discovery of America by Columbus?" Granted that America already existed, was it previously unknown and did Columbus first bring it to light?

Of course not. There were Indians on the spot watching Columbus "discover" America. The ancestors of the Indians (undoubtedly Siberian natives) had discovered the American continents in a process that began some 25,000 years before Columbus—or before the Vikings or Irish or Phoenicians.

What Columbus did was to be the head of the first expedition that brought the American continents to the permanent attention of *Europe*, nothing more. It's just the taken-for-granted racism of Westerners that makes it seem that discoveries only count when Europeans or people of European descent make them.

One would commit the same solecism if one were to speak of the "discovery of a ninth-magnitude star in 1916."

By 1916 the sky had been pretty well photographed down to the ninth magnitude, so that every ninth-magnitude star was on some photographic plate or other. They were all there to be looked at, and it would be difficult to prove that one astronomer had happened to glance at a particular speck of light earlier than another. Even the routine plotting of positions of this speck or that doesn't seem terribly significant.

But then, every star has properties of its own, and a number of them have extraordinary properties. A star with a particular extraordinary property stands out from all the rest and gains an individual distinction. In that sense, the discoverer of that property can almost be said to discover that star—as an individual—and yet that is still loose use of the language.

There is, for instance, a star that has been known as "Barnard's star" since 1916, and one might easily assume that the American astronomer Edward Barnard (see Chapter 11) had discovered the star. In fact, just to show you I'm not assuming a virtue I don't possess, I fell into the trap myself. In my book *Asimov's Biographical Encyclopedia of Science and Technology* (Doubleday, 1972), I say, "In 1916 Barnard discovered a dim star . . ."

In my more recent book *Alpha Centauri, the Nearest Star* (Lothrop, 1976) I was more precise, however, and said, ". . . in 1916, Edward Emerson Barnard noted the . . . rapid proper motion of a star . . . Although many had noted the star before, it was Barnard who first pointed out its proper motion and it is therefore called 'Barnard's star' in his honor."

There! That's the right way to say it.

What's so remarkable about noting a rapid proper motion (that is, the drift, with time, of a star against the general background of the other stars)?

Well, there are something like 130,000 ninth-magnitude stars in the sky, and if you were to try to study each one carefully in order to note some unusual property, you would find the work very time-consuming and tedious and, on the whole, very unrewarding. What usually happens is that something out of the way about some star happens to catch the eye of an observer, more or less by accident, and if he is quick enough to grab onto that and study it further, he may make an interesting discovery.

Thus, Barnard just happened to notice that the star we now call "Barnard's star" was out of position and, by studying older star maps, he noticed there was a progressive change of position. He was then able to demonstrate that the star was moving at a rate of 10.30 seconds of arc per year against the background of the other stars.

This isn't fast in an absolute sense. It means that Barnard's star takes 181 years to move a distance equal to the apparent width of the full Moon. A casual observer looking at the sky and estimating the speed required to cross the Moon in 181 years would be forgiven if he thought that speed was an unbearably slow creep. This happens, however, to be the fastest proper motion known for *any* star. The discovery proved so impressive to astronomers that the star is occasionally referred to as "Barnard's runaway star" or even "Barnard's arrow."

Three years earlier, in 1913, an eleventh-magnitude star had suddenly come into prominence. There are something like 1,200,000 eleventh-magnitude stars in the sky so that to notice something extraordinary about one particular star among them is a markedly greater stroke of fortune than even Barnard was to have three years later. What's more, the property in the case of the eleventh-magnitude star was an even more impressive one than the fast proper motion of Barnard's star.

Robert Innes, a British astronomer in South Africa, noticed that this particular eleventh-magnitude star was not where it was supposed to be and, on checking into the matter, he found that the motion of the star was not progressive but cyclic. He was, in short, observing a parallax. This dim star was describing a tiny ellipse that mirrored the ellipse the Earth was making about the Sun. As the dim star is viewed against the starry background from an Earth that is moving in its orbit, the star is seen from progressively different directions and its apparent position shifts.

The parallax wasn't very large. The dim star shifted from its average position by an extreme of 0.762 seconds of arc, or about 1/2400 the apparent width of the Moon. This happened, however, to be the largest stellar parallax observed up to that time, or, for that matter, since that time.

Let's compare the matter of a large parallax with that of a fast proper motion.

A fast proper motion, such as that of Barnard's star, would indicate a very good chance that the star is quite close to us. There might, however, be stars closer to us, which have smaller proper motions, because they just don't move very fast in an absolute sense or because, while they move fast, that motion is more or less toward us or away from us, and we can only see the crosswise, or transverse, component of that motion.

As a matter of fact, Barnard's star is close to us, but it is not the closest star. Barnard's star is 5.86 light-years away and Alpha Centauri is only 4.40 light-years away, even though the latter has a proper motion only one quarter that of Barnard's star.

Parallax, however, is an unmistakable measure of distance. The larger the parallax, the smaller the distance of the star from ourselves and there are no mitigating circumstances.

Innes's star, if I may call it that for a moment, has a parallax slightly larger than that of Alpha Centauri and it is, therefore, only 4.27 light-years away. No star had ever before, or has ever since, been found to be closer to us than Innes's star (if you don't count the Sun itself, as a wise-guy thirteen-year-old once pointed out to me).

And here is a peculiar coincidence. Innes's star, which is just slightly closer to us than Alpha Centauri is, happens to be located quite close to Alpha Centauri in the sky— about 2° away. Innes's star is just about as far away from us as Alpha Centauri is, and it is in just about the same direction from us as Alpha Centauri is. That means it is not very far from Alpha Centauri in an absolute sense.

Innes's star is, in fact, only 0.16 light-years from Alpha Centauri.

In our section of the Galaxy, the average separation between neighboring stars is 7.6 light-years. To have two stars separated by only 0.16 light-years would be a fairly farfetched coincidence, if there was no connection between the two—if they just happened to be passing another in a near miss at this particular time.

The distance is small enough, however, to make it possible to suppose that there is a gravitational connection. Alpha Centauri itself is a binary star, two stars rotating about a common center of gravity.* The larger, Alpha Centauri A, is almost precisely the twin of our own Sun in all respects. The smaller, Alpha Centauri B, has about seven eighths the mass of our Sun. They circle each other in a period of eighty years.

Could it be that Innes's star is a third member of the group, circling the two chief stars in a vast orbit? In the years since Innes's discovery, Innes's star has been narrowly observed, and from its motion, it does appear possible that it may be circling Alpha Centauri A and B. It is now generally considered part of the Alpha Centauri system and is called Alpha Centauri C.

Because Alpha Centauri C is perceptibly closer to us than the other two stars of the system, it is often called Proxima Centauri, from the Latin word for "nearest." This is not a good name, however, for the nearness of Alpha Centauri C is only a function of its position in its orbit. Eventually, it will move beyond the two chief stars and will then be farther from us than they are.

If Alpha Centauri C is as close to us as the other two stars of the Alpha Centauri system, why is it so dim? Alpha Centauri A, if seen by itself, would be a first-magnitude star; Alpha Centauri B, if seen by itself, almost second-magnitude. Yet Alpha Centauri C, despite its closeness to us, is not even visible to the unaided eye.

To put it another way, Alpha Centauri A is almost exactly as luminous as the Sun, and Alpha Centauri B is 0.28 times as luminous as the Sun. Alpha Centauri C, however, is a "red dwarf" star that is only 0.000053, or 1/19,000, as luminous as the Sun.

Let's indulge in a fantasy. Suppose that our Sun had a far distant red dwarf companion just as its twin Alpha Centauri A does.

Impossible, you say? We'd spot it in a minute? Well, let's see.

* See "Planet of the Double Sun," in my Doubleday collection *Asimov on Astronomy*.

Let's imagine that such a star actually exists. We will call it Proxima, for if it existed it would truly, and permanently, be the nearest star to ourselves except for the Sun. What would it be like?

Suppose we make it just as far from the Sun as Alpha Centauri C is from Alpha Centauri A—0.16 light years. That would mean Proxima was 1,500,000,000,000 kilometers (940,000,000,000 miles) from the Sun or 205 times as far from the Sun as the planet Pluto is, even at the latter's farthest point of recession.

Imagine ourselves observing the skies from a point near our hypothetical Proxima. As far as the stars were concerned—that is, all but one—there would be virtually no change from the appearance as seen from Earth. On the stellar scale, a shift in viewing position of 0.16 light-years is insignificant.

That "all but one" refers to the Sun itself, of course. The Sun would not be an object of overwhelming brilliance, but it would still be brighter than any other star. It would have a magnitude of just about −7, which would make it 150 times as bright as the next brightest star, Sirius.

At Proxima's great distance from the Sun, it would take an ordinary planet just about 1,000,000 years to make one circuit of the Sun. However, Proxima (if we suppose it to be as massive as Alpha Centauri C) would have a mass about one fourth that of the Sun. The gravitational attraction would depend upon the sum of their masses so Proxima would move a bit faster than a planet would and would complete its circle in about 900,000 years.

And how would Proxima appear to us here on Earth?

Not very bright at all. It would be visible to the unaided eye, of course, but not overwhelmingly so. It would have a magnitude of about 3.5. There would be about two hundred stars in the sky brighter than Proxima.

Nor would it look remarkable through a telescope. Proxima would be a small star (if it is like Alpha Centauri C), only three times the diameter of Jupiter, and it would be a sixth of a light-year distant. Its apparent diameter at that distance would be only about 0.01 second of arc and it would look like nothing more than a point of light in any of our telescopes. If it could be expanded into a visible orb that would attract attention, of course, and give it away—but it couldn't be.

How about its motion? After all it would be moving around the Sun and therefore should be moving in our sky. Would that be the tip-off?

Its average speed across the sky would be about 1.44 seconds of arc per year. At the end of each year it would be 1.44 seconds of arc farther east than it was at the beginning.

That's not much. Barnard's star has a proper motion that is over seven times as great. Alpha Centauri has a proper motion that is 2.5 times as great.

Why is this? If a star that is near to us is apt to have a large proper motion, why wouldn't Proxima, which would be *so* near to us, have an overwhelming proper motion?

The proper motion of the stars is the result of the fact that all the stars in our neighborhood are whipping around the galactic center at great speeds. Our Sun moves in its galactic orbit at a speed of 250 kilometers (155 miles) per second. If all the other stars were doing exactly the same, they would all seem motionless relative to ourselves.

The other stars are not, however, doing quite the same. Some are nearer the galactic center and some farther. While the Sun has a nearly circular orbit, some of the other stars have elliptical orbits. Some of the stars have orbits in planes that are tipped relative to the Sun's. All these factors—distance, ellipticity, orbital tilt—introduce differences in speed. It means that most of the stars have considerable velocities relative to the Sun and a few may have relative velocities of 200 kilometers (124 miles) per second. If these are close enough to the Sun and if the relative motion is crosswise to our line of sight, then this is reflected in a large proper motion.

In the case of Proxima, however, it would be moving

around the galactic center along with the Sun itself, and in exact step. It would seem motionless to us as far as the galactic orbit were concerned.

What speed would exist relative to ourselves would be the speed of Proxima's orbital motion around the Sun and this would be a real creep. It would be moving at a speed of only 0.33 kilometers (0.20 miles) per second. Even Proxima's relative closeness to ourselves couldn't make much out of that.

To be sure, Proxima's orbital motion would be a very uneven one. For a half a year, it would move west to east, starting from zero speed, increasing to a maximum after three months, and falling to zero after six months. It would then reverse direction and move east to west for another half year, starting from zero, increasing to a considerably smaller maximum after three months, and falling to zero after six months—then repeating. It would take two steps forward, one step back, two steps forward, one step back, and so on.

The reason for this would be that when the Earth is on the opposite side of the Sun from Proxima, it would be moving in the direction opposite to that of Proxima, and the two speeds would be summed, so that Proxima would seem to move particularly rapidly.

When Earth would make the turn and come over on the same side of the Sun as Proxima, it would overtake the much more slowly moving Proxima. Proxima therefore would appear to be going backward, though at a slower rate than it had earlier been going forward.

As Proxima circled the sky in 900,000 years, it would appear to make 900,000 tiny loops, which would really amount to a very large parallax of about 20 seconds of arc—and that would prove its extreme closeness to our Sun.

But would those loops be noticed?

If someone bothered to compare Proxima's position in the sky with that in charts taken several decades ago, he would come up with an average proper motion of 1.44

seconds of arc per year, and it might never occur to him to examine that proper motion closely for periodic deviations.

Then, too, suppose Proxima's orbit were not in the general plane of the planetary orbits, as I have been assuming. There is no reason it should be. It might be moving in an orbit well tipped to the plane of Earth's orbit and it might be located far outside the zodiac. Proxima might then be all the more apt to go unnoticed.

Since the planets all move through the constellations of the zodiac, those constellations get a great deal more routine attention than other constellations do. In addition, the effect of Earth's motion on Proxima, if it were in an orbit highly inclined to our own, would be to make it move not so much in loops as in very shallow tiny waves—900,000 to each complete orbit. The waves would be less likely to be noticed casually than the loops would be, in my opinion.

But suppose someone were, quite by accident, to notice that Proxima were moving in loops or waves. Would the conclusion that it was a close companion of our star be inevitable?

Perhaps not.

Suppose the astronomer studying the motion were convinced that, like all other stars, Proxima was far away from us—say, several light-years away at the least. Might it not be possible that he would diagnose the looping or waving as the result of the existence of a companion to Proxima—a white dwarf, perhaps, or a massive planet?

Sirius and Procyon move with a kind of wavy motion and each has a white dwarf companion. Barnard's star moves with a small wobble and so does 61 Cygni and they are thought to have massive planets in consequence.

To be sure, one would have to deduce that Proxima's companion, whatever it were, would be circling Proxima in a period of exactly one year, but that could be put down to coincidence perhaps.

Even the fact that Proxima would be seen spectroscopically to be neither approaching us nor receding from us at any perceptible speed and, thanks to its small average proper motion, was keeping step with us almost exactly in our swim around the galactic center, might be put down to coincidence, if the matter happened to be observed at all (which perhaps wouldn't be likely).

There would be another possible giveaway.

Proxima would not be revolving about the sun, really. It and the Sun would be revolving about a common center of gravity.

Each planet revolves about the common center of gravity of itself and the Sun, but the mass of the Sun is so overwhelming that there isn't very much difference between the center of the Sun and the center of the Sun-planet center of gravity. The difference amounts to 460,000 kilometers (285,000 miles) in the case of Jupiter, but even that isn't very much.

Proxima, however, would be one fourth the mass of the Sun and is one sixth of a light-year away. That means that the center of gravity of the Sun-Proxima system would be on a line connecting the centers of the two bodies, one fourth of the way from the Sun to Proxima and therefore 375,000,000,000 kilometers (235,000,000,000 miles) from the Sun.

The Sun and its entire family of orbiting objects—planets, satellites, asteroids, comets, debris—would be circling that point once every 900,000 years as they move in their 200,000,000-year circuit of the galactic center.

This means that the Sun and its family would move in an orbit about the galactic center made up of 220 shallow waves in every circuit. To us on Earth it would seem as though the rest of the Galaxy were waving slightly.

No one star, of course, is likely to reveal this apparent Galactic wave, since each will be having some irregularities of motion of its own, most likely. If the motions of many stars are averaged in a statistical manner, however, that average would show the waves. Rather than suppose that the whole Galaxy waves, we would surely conclude that the Sun were moving in a wavy orbit and deduce the presence of Proxima from that.

The wave would be a long shallow one, however, very long in comparison with the history of modern astronomy and very shallow in comparison to stellar distances. We have been observing stellar positions and motions with considerable accuracy for, say, two hundred years, or 1/4500 the length of one of those waves. That's not long enough for detection, I think.

Two more points.

It is suspected that the Sun has a shell of comets circling it at a distance of a light-year or two, far beyond even the orbit of the hypothetical Proxima, and that some of them are perturbed by nearby stars and, as a consequence, drop into the inner solar system and skim the Sun.†

If Proxima existed, would it not have a much greater perturbing effect on the comet shell than any of the outer stars would? Could one deduce from the orbits of the long-period comets we detect that Proxima exists or does not exist? If it exists, could we calculate where it might be now?

For that matter, would Proxima exert measurable perturbations on the orbits of the outer planets?

I don't know.

The second point arises out of the fact that I have been assuming that Proxima would follow a circular orbit about the Sun, or, which is the same thing, that the two stars would circle a common center of gravity in such a way as to be separated by a constant distance.

It's rather more likely that Proxima's orbit would be markedly elliptical and that the separation between the two stars would decrease and increase in a very slow rhythm.

It may be that the luck of the game would put humanity

† See "Steppingstones to the Stars," in my Doubleday collection *Asimov on Astronomy.*

and its advanced astronomic instruments on Earth at a time when Proxima was just about at apastron—that is, at maximum distance from the Sun. Perhaps, over the next few hundred thousand years, Proxima would be approaching the Sun and ourselves steadily—growing steadily brighter and moving across the sky at a steadily increasing pace. Eventually, it would be impossible to miss.

And during all the early Stone Age centuries, Proxima would have been receding steadily from the Sun, growing dimmer and moving more slowly until it became easy to miss.

Perhaps if we had come on to the scene a couple of hundred thousand years later—or a couple of hundred thousand years earlier—

So here it is in a nutshell.

It isn't likely that the Sun has a distant red dwarf companion—but it just might.

If it does have a distant red dwarf companion, it isn't likely that we would be missing the fact now in the twentieth century—but we just might.

Proxima might be out there and we might be overlooking it, chiefly because it occurs to no one that it's there and no one's really searching for it.

—Or is there some aspect of the situation that I've missed, and is there some giveaway that would so surely be spotted that in the absence of such spotting, we can be certain that nothing like Proxima exists?

[*Note:* Actually, just about all visible stars have been tested for parallax, I discovered, since I wrote the above chapter. Proxima is not likely to have been overlooked, therefore, and does not exist if it is a star with the properties of Alpha Centauri C. What if, however, it is a somewhat smaller and dimmer star, and possibly farther from the Sun? In that case, Proxima could not yet be ruled out.]

F UNIVERSE

14. Fifty Million Big Brothers

In preparing the index list of my twenty years of essays, which you'll find in the introduction to this book, I became particularly aware that seven of my early essays were never included in any of the collections I have periodically published with Doubleday.

In each case, it was because, for one reason or another, I was dissatisfied with the essay.

Yet it offends my sense of neatness to leave matters so. If a particular essay was, in my own mind, unsatisfactory and if the subject matter is not completely unimportant, then, after a suitable interval, ought I not to try again, doing better this time?

The seventh and last of my uncollected essays was "Who's Out There?" which appeared in the September 1963 issue of *F & SF*, a little over fifteen years ago. It dealt with the possibility of intelligent life elsewhere in the Universe, and this is a subject which has not faded away in the years that have passed. Indeed NASA is proposing to spend five years and $20,000,000 to search the heavens for signals that are neither perfectly regular nor perfectly random and that many therefore be of intelligent origin.

What are NASA's chances at success?

May I now tackle this subject once again?

In order to come up with an answer to my question as to NASA's chances, I will have to make several assumptions. The two most basic (and perhaps quite debatable) assumptions are these:

1. The only life that exists is life-as-we-know-it—that is, life based on nucleic acids and proteins doing their thing against a water background. This is not really a very restrictive assumption. Our experience on Earth shows that a wide variety of life has existed on this one planet—tens of millions of species with a bewildering array of enormous superficial differences—all of which are basically similar on the biochemical level. Undoubtedly there is sufficient room for variety to allow an equal array of species on every planet of every star in the Universe, with no exact duplications anywhere.

But why can't we allow variations in the basic pattern, too? A background of liquid ammonia, or liquid silicones, or liquid hydrogen? Complex molecules of fluorocarbons or silicates? Or, for that matter, why not gaseous life, solid life, nuclear-reaction life, or pure mind-life?

We can postulate these if we wish, but there is no single scrap of evidence for the existence of any of these things, and to speculate in the absence of any evidence at all is to produce something so undisciplined that any answer is possible. And where any answer is possible, all answers are meaningless.

The value of assumption No. 1 is that we can eliminate from consideration any environment that is incompatible with our kind of life. This allows us to eliminate many environments for known reasons and therefore tends to give our final conclusions meaning.

2. The situation on Earth is average. It has in no way followed an unlikely course, either by taking advantage of an unbelievably lucky break or by falling victim to an unbelievably unlucky break.

Mind you, this, too, is an assumption. We have no reason to think that the situation on Earth is average, but no reason to think that it is not average, either. If it is average, however, we can make certain estimates. If it is not, then we must be at such a loss to decide in what way it is not average and to what an extent—that again we can decide anything and, therefore, nothing.

* * *

Now we are ready to begin.

By assumption No. 2 we can decide that life must begin, as it did here, somewhere in the neighborhood of a star which can supply the necessary energy for the formation and maintenance of life. By assumption No. 1 we decide that the star in question must be something like the Sun in nature, for only in that case can our form of life be maintained.

The Sun is a star of moderate size. There are dim, cool stars with masses as little as one fiftieth that of the Sun, and there are brilliant, hot stars with masses as much as fifty times that of the Sun.

A star with a small mass delivers very little energy compared to that delivered by the Sun. For a planet to receive enough energy from a small star for the needs of life, it would have to be in a close orbit, circling the star at a distance of perhaps as little as 150,000 kilometers (90,000 miles).

While energy delivered varies inversely as the square of the distance, the tidal effect varies inversely as the cube of the distance. This means that by the time a planet has approached its star closely enough to get the energy it needs, it is getting far too much tidal effect. The planet's rotation will be slowed until it faces one side always to the star—ending up with a hot side, a cold side, and, probably, not much in the way of an atmosphere.

A star with a large mass has a short lifetime on the main sequence between the time it first forms and the time it expands to a red giant. Our experience on Earth is that it takes a long time for an intelligent species to develop, and if this is so more or less everywhere (assumption No. 2), then it is useless to expect intelligent life in the neighborhood of large, hot stars.

We therefore end up looking for Sunlike stars, those with masses not less than 0.4 and not more than 1.5 times that of the Sun.

Our first question, then, is, How many Sunlike stars are there in the Universe?

It is hard to answer this question because we don't really

know how many stars of all kinds there are in the Universe. The stars are collected into galaxies, and our telescopes show us many millions of galaxies but there are undoubtedly many millions of others that we do not see. The most liberal estimate I have seen of the total number of galaxies in the Universe is 100,000,000,000, in which case the total number of stars must be of the order of magnitude of thousands of billions of billions, but with a wide possible variation to allow for our very uncertain knowledge of the actual total number of galaxies.

To get a more meaningful number, let's limit ourselves to our own Milky Way Galaxy. For one thing, the possible intelligent life-forms in other galaxies are at a distance from us of anywhere from millions to billions of light-years. Intelligent life-forms within our own Galaxy, however, are, at most, 150,000 light-years away. It is reasonable to suppose, then, that the intelligent life in our own Galaxy is much more likely to be of importance to us than intelligent life elsewhere.

Besides, any conclusions we can come to about our own Galaxy will hold also, on the average, for all the other galaxies, by a natural extension of assumption No. 2.

If we concentrate on our own Galaxy, then, its total mass, according to the latest estimate I have seen, is 200,000,000,000 times that of our Sun. A third of that is liable to be in the form of dust and gas which means that the starry portion of the Galaxy has a mass of about 140,000,000,000 Suns. The mass of the Sunlike stars is only about one tenth of the total mass, or 14,000,000,000 Suns.

Since the Sunlike stars are individually about equal to the Sun in mass, there are 14,000,000,000 Sunlike stars in the Galaxy.

One thing I didn't do in 1963 was to make allowance for the position of the Sunlike stars in the Galaxy, since at that time all parts of the Galaxy seemed equally hospitable (or inhospitable) to life.

We no longer think so.

In 1963 the quasars, just discovered, were still a complete mystery. That mystery has by no means been resolved even now, but there is a growing concensus that the quasars are galaxies with extremely active and brilliant centers. They are so far away that the galaxies themselves cannot be seen even in our best telescopes, but those blazing centers show up for all the world like faint stars (and would be interpreted as nothing more than such, were it not for the telltale evidence of microwave emission and enormous red shifts).

But if quasars are galactic centers blazing with the light of a hundred ordinary galaxies (as they must be, to be visible at the enormous distances of from 1,000,000,000 to 10,000,000,000 light-years), something very unusual and violent must be going on in those centers.

As a matter of fact, galactic centers everywhere have come to seem wild places. Our own Galaxy, for instance, has a very active microwave source confined to a very small area in the sky, and a dramatic explanation of this would be to suppose that there was a monster "black hole" at the galactic center, one with a mass equal to 100,000,000 Suns and therefore 1/2000 the mass of the entire Galaxy. It is growing, naturally, and may be gulping down whole stars, when such stars have motions that spiral them in too closely to the all-embracing maw of the black hole.

It may be that black holes naturally form wherever stars are densely packed, as at the centers of galaxies or, to a lesser extent, at the centers of globular clusters. It has even been suggested that galaxies form in the first place about black holes, that each galaxy is an accretion disk about a black hole.

Black holes or not, the increasing evidence of violent activity in the centers of galaxies, including our own, would make it appear that galactic nuclei are inhospitable to life. The radiation level would be too high.

This means that life would be possible only in the quiet suburban volumes of the Galaxy—out in the spiral arms where our own Sun is located. Since some 90 per cent of the mass of the Galaxy is located in its nucleus and only

10 per cent in the spiral arms, we must consider that the number of potentially life-supporting Sunlike stars is only one tenth the total number, or 1,400,000,000.

Naturally, a star cannot support life unless there is a planet on which the life can originate. By current theories of planetary origin from a condensing cloud of dust and gas, it would appear that, in the process of star formation, planets are also formed in the outskirts of the cloud.

If every cloud of dust and gas condensed into a single star, that would be that. However, it is quite common for a cloud to collapse into two stars, forming a binary. This binary may be associated with another star or two or, for that matter, with another binary or two. Binaries are, however, invariably very widely separated from any associated stars. From the standpoint of planetary formation, we therefore need not consider any association more complex than that of a binary.

When the stars of a binary are themselves separated by a respectable distance, each of the stars may have a planetary system unaffected to any great degree by the other. If the binaries are close together, however, any planets that form about one are liable to have such unstable orbits they will not be long-enduring, and any formed about both treated as a gravitational point would be so far distant from both stars as to receive insufficient energy for life.

Perhaps half the stars in existence are members of binaries, and perhaps half of those are members of close binaries which if they have planets at all, do not have the kind of planets that are compatible with life.

Therefore, we can conclude that only about three quarters of the potentially life-supporting Sunlike stars have potentially life-supporting planetary systems. The total number of potentially life-supporting planetary systems in our Galaxy is then just about 1,000,000,000.

A planetary system may be potentially life-supporting and yet may not have a planet that can actually support life.

Our own planetary system is obviously life-supporting and yet only on Earth is there life. There is certainly none on the Moon, for we have looked. There is almost certainly none on Mars, for our machines have looked. The environments on the remaining planets are sufficiently hostile (by assumption No. 1) to make it seem quite likely that they do not contain life, either.

Furthermore, Earth itself would easily have been non-habitable if it were somewhat smaller or larger than it was, or somewhat closer to or farther from the Sun, or if its orbit about the Sun were a little more eccentric, or its period of rotation were a little longer, or its axial tip were a little more pronounced.

In this respect, then, assumption No. 2, that Earth is average, cannot possibly be maintained. Every significant change in Earth's size, structure, location or motion, would seem to be for the worse. Granted that this may be only a matter of appearance, since life is adapted to the situation on Earth exactly as it is, yet, considering the fragility of the nucleic acid/protein system it is hard to believe that there isn't considerable truth beyond the appearance, too. After all, Venus, Mars, and the Moon, which are worlds that are not enormously different from Earth, do not carry life.

With Earth not average, but at a favorable extreme, could we suppose that every potentially life-supporting planetary system would, like the solar system, have an Earthlike planet? That would be a height of inadmissible optimism.

It would, on the other hand, be the depth of inadmissible pessimism to suppose that no potentially life-bearing planet would appear anywhere else and that only here on Earth itself, in all the Galaxy, would we encounter a planet that had the good fortune to hit all the requirements on the bull's-eye (or at least, as sufficiently near the bull's-eye as made no difference).

The truth is most likely to be somewhere between 0 and 1 Earthlike planet per planetary system, but where between? There is absolutely no way of telling. We can only guess, and my own guess is one Earthlike planet for every ten planetary systems. We can call this assumption No. 3 in my line of argument, though it is a far less all-embracing one than the first two.

If this is true, then the number of Earthlike planets suitable for life in our Galaxy comes to $1,000,000,000 \times 0.1$, or $100,000,000$.

A planet may be suitable for life and yet not bear life. It is very tempting to think of life as something miraculous and the product of divine creation. Even those relatively few people who are willing to suppose life to be the result of an accidental concatenation of atoms, can be so overwhelmed by the utter complexity and versatility of present life as to assume that the probabilities of such an accidental origin are incredibly low. They might even suspect that however many Earthlike planets there might be, it would be on Earth only that life would occur.

This, too, strikes me as the inadmissible depths of pessimism, and here, at least, we have observational evidence to demonstrate its inadmissibility.

Beginning in 1955, chemists have experimented with mixtures of simple chemicals of the sort there is every reason to suppose existed on Earth in primordial times prior to the advent of life. If this mixture were subjected to the kind of energy to which the primordial Earth was subjected—from the Sun, from volcanic heat, from lightning, from radioactivity—then in a matter of days or weeks, more complicated chemicals would be built up. These complicated chemicals could be used as starting points and then still more complicated chemicals would be built up.

Even the most complicated chemical formed in this fashion is at an enormous distance from even the simplest recognized form of life, but they point in the right direction. Amino acids are formed, nucleotides, adenosine triphosphate, even proteinlike molecules. If we can do that in small vessels in weeks, what can be done in an entire ocean in a million years?

Nor is this merely the unconscious predilection of scientists who might unconsciously arrange an experiment in such a way as to insure the answer that would be most thrilling. In the 1970s moderately complex organic compounds were discovered in freshly fallen meteorites of the carbonaceous chondrite variety—compounds that were clearly formed in the absence of life and yet are pointed in the right direction even though no scientist was around to do the pointing.

In fact, even in the vast dust clouds between the stars, atoms come together in random collisions and form molecules containing up to nine atoms (as far as we have detected to this point), and these, too, point dimly in the direction of life.

We have every reason to think, then, that given Earthlike conditions and an Earthlike chemical constitution, living things are bound to occur eventually. Far from life's origin being a miracle, it would be rather miraculous for it not to come to be.

But wait! How long is "eventually"?

On Earth, the oldest fossils (as we ordinarily think of fossils) are some 600,000,000 years old, but Earth itself, as a solid body, is some 4,600,000,000 years old. From the first 4,000,000,000 years, there are no fossils. Did it then take that long for life to form on Earth even with our planet's apparently ideal conditions for life? Would not even a tiny veering away from the ideal lengthen the time required until life never begins at all?

No, for this is an underestimate of the age of life on Earth. The fossils that first appear in rocks that are 600,000,000 years old are the fossils of very complex organisms—organisms large enough to be seen with the unaided eye, easily recognized as life-forms, and often with shells and other hard parts that easily fossilize. Before these developed, there must have been a long history of smaller and simpler organisms, perhaps one-celled in nature, the traces of which are much fainter and more subtle than that of ordinary fossils.

The faint and subtle traces have been found, and micro-organisms have been traced back in rocks that are well over 3,000,000,000 years old. When Earth was only 1,000,000,000 years old, it was teeming with life and it is very possible that life formed no later than 500,000,000 years after the Earth had formed.

The average period of time during which a Sunlike star remains on the main sequence is about 10,000,000,000 years.

Before a star arrives on the main sequence, it is merely a mass of condensing dust and gas, while the planets themselves are merely coalescing bodies. There is no life then.

After a star leaves the main sequence, it expands into a red giant, frying to death any life-bearing planet that circles it.

The average period of time during which an Earthlike planet can support life, then, is 10,000,000,000 years.

The various Earthlike planets in the Galaxy are bound to be of different ages, since stars have been forming all through the history of the Galaxy. Some are forming right now, and some will be forming 1,000,000,000 years from now.

If we assume that stars and planets have been forming in the Galaxy at a constant rate (probably not quite true), we can say that 5 per cent of the Earthlike planets have expended less than 5 per cent of their lifetimes by now; 15 per cent have expended less than 15 per cent of their lifetimes; 50 percent have expended less than 50 per cent of their lifetimes; and so on.

If life appeared on Earth 500,000,000 years after its formation and if this is an average event (by assumption No. 2) and likely to happen, give or take a few million years, on all Earthlike planets, then any Earthlike planet older than 500,000,000 years would have life upon it in some stage of development.

Half a billion years is 5 per cent of a life-supporting planetary lifetime, and only 5 per cent of such planets are therefore less than 500,000,000 years old. That means that 95 per cent, or 95,000,000, of all the Earthlike planets suitable for life possess life, while the remaining 5,000,000 planets are crawling with chemicals on the way to life.

It may be that 95,000,000 independent life-systems in our own Galaxy sounds like a great deal, but it means that only one out of every 1500 stars in the Galaxy shines down on a life-bearing planet.

Life in itself is something, but it is not enough. What we are talking about is intelligent life.

On how many life-bearing planets does intelligence develop? Specifically, on how many life-bearing planets does a species develop which is capable of constructing a technological civilization?

If we think about it, it is bound to take a long time. Intelligence is a valuable thing, but it is not usually the key to survival. Sheer fecundity is usually what counts. The intelligent gorilla doesn't do as well as the less intelligent but more fecund rat, which doesn't do as well as the still less intelligent but still more fecund cockroach, which doesn't do as well as the minimally intelligent but maximally fecund bacterium.

Therefore, we might expect that evolution will curve in the direction of fecundity rather than intelligence. If intelligence does develop in some odd byway, it is only in combination with a few other things, like hands and good vision, that it can reach the point where it can begin to make up for low fecundity. If intelligence reaches the point where its owner is capable of changing the environment to suit himself, then and only then does it have a chance to become the overwhelming factor. The early hominids just managed to squeak past that critical point and perhaps only with the development of fire and the stone-tipped spear did intelligence begin to show what it could do.

It took 4,600,000,000 years for intelligence to pass the critical point on Earth and for a technological civilization to become possible. That means roughly 50 per cent of the lifetime of Earth as a habitable planet.

If we go by assumption No. 2 and suppose that this has

happened, give or take a few hundred million years, on other life-bearing planets as well, then we can conclude that on half the life-bearing planets a species has arisen intelligent enough to establish a technological civilization.

Since we have calculated that there are 100,000,000 planets that bear life or are about to bear life and since half of them have reached or passed the midpoint of their lives, assuming a constant rate of formation of planetary system, there has been time for no less than 50,000,000 technological civilizations to have come into being in our Galaxy.

Our own technological civilization has only been in high gear since the 1770s, with the invention of a practical steam engine. Considering how far we have come in two hundred years, consider how far we might come in another thousand years. We would by then surely have a technology far beyond the present.

A thousand years, however, is only one five-millionth of a planetary lifetime. All but ten of the technological civilizations would be more than one five-millionth of a planetary lifetime older than we are and we might as well say that there are 50,000,000 technological civilizations in the Galaxy that have come into being long enough in the past to be, at present, far more advanced than we are.

We can conclude, then, on the basis of experiment, observation, and three assumptions, that we ourselves, who are just emerging from childhood, are trying to contact 50,000,000 big brothers out there.

—And the discussion will be continued in the next chapter.

15. Where Is Everybody?

My beautiful, blue-eyed, blonde-haired daughter, who just yesterday (as I write this) received her college diploma, with myself waving and shrieking madly from the stands, is a constant source of stories concerning the reactions to her name.

Recently, she was stuck in an airplane that was delayed over two hours before taking off. She couldn't get off to call me about the delay and she knew I was expecting her to arrive at her destination at a certain time and to call me, at that time, to the effect that she had arrived safely. She knew well that I would be upset at the unexplained delay and she felt annoyed at this.

A kindly businessman in the next seat went out of his way to occupy her mind and keep her calm by engaging her in lively conversation. Finally, he asked her what this father of hers (about whose peace of mind she seemed so concerned) did for a living.

"He's a writer," said Robyn.

"What does he write?" asked the businessman.

"Well, he's best known for his science fiction."

"You mean the kind of stuff Asimov writes?"

"Exactly. In fact, he *is* Asimov."

Sensation!

When things calmed down, the businessman said, "Tell me something. How did your father know, when he began to write science fiction, that he was going to be successful?"

It was a logical question from the business point of

view. After all, if an enterprise doesn't promise success, you don't undertake it.

And, alas, that is the same attitude that many "practical" men take toward a scientific project that requires vision extending an inch or two past the tip of one's nose. For instance, the notion of a search for extraterrestrial intelligence seems like a boondoggle to them even though, as I explained in the previous chapter, one can reason out, legitimately enough, the good possibility that as many as 50,000,000 technological civilizations have gotten their start here and there in our Galaxy.

To be sure, my reasoning in the preceding essay is not sufficient, in itself, to allow me to wax merrily sarcastic at the expense of those who doubt a search for extraterrestrial intelligence will succeed. There is one question that remains to be asked.

Here it is: "Where is everybody?"

If there are 50,000,000 technological civilizations more advanced than ourselves, why haven't we heard from them? Why haven't they visited us? Why is everything so quiet?

Let's consider the possible answers to the question.

1. *There may be some serious mistake in my reasoning in the previous essay and it may be that there are, after all, no technological civilizations anywhere but on Earth.*

But where can the mistake be? We *know* there are a vast number of stars and that a substantial portion of them are Sunlike. We have good reason to believe that planetary systems are common and, if so, there should be, just by sheer chance, a goodly number of Earthlike planets. We have good reason to believe that life is an all but inevitable development, and in that case why shouldn't intelligence and civilization be found as a reasonably common development of life?

Where, then, could we have gone wrong?

If despite all these sound arguments, a civilization exists only on the Earth, then there must be something very unusual about the Earth that we haven't taken into account— some characteristic we are not likely to find on other planets that, superficially, may seem Earthlike. But if so, what can it be?

What about the Moon?

It is enormous. It has 1/81 the mass of the planet it circles. There are five other planets with satellites, but in every case, those satellites are tiny compared to their primary.*

The Earth-Moon system is virtually a double planet. It is as though it formed out of the original nebula about a double nucleus, while all the other planets formed about a single nucleus with some inconsiderable remnants clotting on the outskirts.

Is the double-nucleus origin so rare that it virtually never happens and does it contribute, somehow, to making Earth a life-bearing planet?

After all, Venus has just about the same size, mass, density, and composition of the Earth, yet it is totally hostile to life as we know it—and it has no satellite.

Does this mean that all the hundred million Earthlike planets we postulated in the last essay are really Venuslike planets and that only our Earth, alone among them all, is Earthlike and can support life? If so, our previous reasoning falls to the ground.

Yet why should we think that the mere formation of the Moon can turn a Venus into an Earth? In fact, it makes much more sense to think of Venus as hostile and Earth as benign because Venus is a bit too close to the Sun and that the difference has nothing to do with the Moon.

It is not even certain that the Moon was formed out of the same dust cloud from which Earth was formed. Earth may well have begun life as a single planet. The Moon may have formed as a single planet, too, but may have been captured by the Earth—perhaps comparatively late in Earth's life.

---

* Since this was written, Pluto has been found to have a satellite, Charon, that is 1/10 its own mass. Both Pluto and Charon are quite small, however.

After all, the Moon has only three fifths the density of Earth and lacks a metal core. It is much more like Mars in these respects than like Earth, so that perhaps it was formed out of the Martian portion of the original cloud.

Then, too, the Moon lacks the more volatile elements, and bits of glassy materials, formed by rocky substances that have melted and re-solidified, are common on it, though rare on the Earth. It may have been subjected to Mercurian heat at some time in the past.

Perhaps the Moon had an eccentric orbit, originally, that brought it to a Mercurian approach to the Sun at perihelion, and a Martian recession at aphelion. Then, at some time, thanks to some tricky bit of celestial mechanics, Earth captured it.

Would that have made a difference? Is there any sharp revolution in Earth's development that took place late in its history?

How about the development of land life? Life in the ocean began perhaps 500,000,000 years after the Earth was formed, but life on land appeared 4,200,000,000 years after. Why the 3,700,000,000-year wait?

Perhaps ocean tides are the secret to land life. The periodic progression of water up a shore and then down again would carry life with it. It would leave pools behind in which some forms of life could flourish. There would be water-soaked sands that could become hospitable to life. Adaptations would make it possible for life-forms to withstand limited amounts of drying between one high tide and the next. These would creep farther and farther up the shore until finally life was possible without any actual immersion in water at any time.

A Moonless Earth, however, would have only the small tides produced by the Sun—one third the amplitude of the tides we now have. These might have been insufficient to get land life started.

If, however, the Moon were captured 400,000,000 years ago, tides would suddenly become markedly greater in ude. In fact, they must have been greater in amplitude then than they are now. Tidal action is slowing the Earth's rotation and driving the Moon farther from the Earth. Working backward, we can show that 400,000,000 years ago, the day was only about 21.8 hours long and the Moon was only 320,000 kilometers (200,000 miles) from Earth.

The closer Moon produced tides 1 2/3 as great in amplitude as it does today, and these would move up and down the shores at a speed 10 per cent greater than at present, thanks to the shorter day.

We might conclude, then, that it was the tides produced by a large captured satellite that made land life possible.

We can also argue that it is only on land that life can develop the long-distance vision and the manipulative organs that make a large brain and a high intelligence possible.† What's more it is only on land and in a free-oxygen atmosphere that fire is possible, and it is the control of fire that marks the beginning of a high technology.

Consequently, unless an Earthlike planet captures a large satellite (and the chances of that are so small that no one has yet figured out how Earth could have managed to do so), the best a planet can hope for is sea life. That would mean that although life-bearing planets are as common as I estimated in the previous chapter, *civilization*-bearing planets would be extremely rare and Earth might even be the only one in the Galaxy.

This, however, is not a compelling argument, either. We are not at all sure the Moon was captured. What's more, tides may not be the crucial factor in the development of land life. It is much more persuasive to argue that the land became habitable only after enough free oxygen had been pumped into the atmosphere by photosynthetic organisms to make an ozone layer possible in the upper atmosphere so that the deadly ultraviolet rays of the Sun could be blocked off. Until that happened, sea life was protected

---

† Dolphins may be intelligent, but they had land-living ancestors. The most intelligent creatures whose ancestors have always been sea-living are the octopuses.

by the uppermost layer of water, but land life had no protection at all.

Consequently, I don't believe the Moon was crucial in either the formation of life generally or the formation of land life in particular. What's more, I can't really see that Earth is intrinsically remarkable in any way other than in the possession of an unbelievably large satellite, so we must *still* maintain that 50,000,000 extraterrestrial civilizations have come into being in our Galaxy.

2. *Wait! They have "come into being"—who says that they stay in being?*

Suppose that each civilization that comes into being lasts for only a comparatively short time and then comes to an end. That would mean that if we could examine all the Earthlike planets in the Galaxy, we might find that on a large number of them civilizations have not yet arisen and that on an even larger number civilizations have arisen and have already become extinct. On only a few planets would we arrive at a time when a civilization had arisen so recently that it had not yet had time to become extinct.

The briefer the duration of civilizations on the average, the less likely we are to encounter a world on which civilization has come and not yet gone and the fewer civilizations *in being* there will be at this moment—or at any given moment in the history of the Galaxy.

Might it be, then, that civilizations are self-limiting and that the reason none of them have visited us is that none of them have remained in existence long enough to make the visit possible?

But why should a species intelligent enough to form a civilization not be intelligent enough to survive?

Judging from the experience of the one species we know that is capable of civilization—our own—we might argue that the mere possession of high intelligence means there is foresight and memory. We can foresee the possibility of discomfort and deprivation and we can remember having been subjected to discomfort and deprivation. We therefore compete with far more violence and persistence
s intelligent species do for desired objects that are

scarce and, if we lose, we are much more likely than they to seek revenge. This is not because we are worse than other species—merely more intelligent.

Contentiousness and violence, in other words, come with the territory and, as intelligence multiplies the force at the disposal of an intelligent species, the violence grows steadily more deadly. Finally, the time comes when the weapons of the intelligent species are so powerful and destructive that they outstrip the capacity of the species to recover and rebuild—and the civilization automatically comes to an end. Even if, long after, the species rebuilds or an entirely new intelligent species arises, suicide again comes as the end.

Homo sapiens appeared on this planet about 600,000 years ago and may be, as a civilization, on the point of extinction. If we assume that the Sun will remain on the main sequence and be capable of maintaining Earth in its habitable state for 12,000,000,000 years, then humanity, existing for 600,000 years, will have maintained itself for 1/20,000 the lifetime of the Earth.

If there are 100,000,000 Earthlike planets in the Galaxy, as I suggested in the previous chapter, and if our history is typical of intelligent species generally, then on only 5000 of these planets is there an intelligent species capable of constructing a civilization which has not yet had time to commit suicide—and none of these are more advanced than we or they would already be extinct.

By this line of argument, then, it is no wonder that we have never been visited by extraterrestrial intelligences.

We can, with particularly deadly significance, use an argument in reverse: (A) Many extraterrestrial intelligences must have established civilization; (B) none have come to visit us; (C) therefore, all have destroyed themselves before they gained the capacity to come visiting, and we ourselves have no chance of avoiding self-destruction, either.

—Yet how can we be sure that our own experience is typical of tens of millions of others? Isn't it possible that at least some intelligent species may use their collective intelligence to foresee the suicide and modify their behavior accordingly? Isn't it even possible that we ourselves may yet do so?

Suppose, for instance, we estimate that there is one chance in ten that we will co-operate, as a species, in the face of destruction, try to solve our problems, control our population, conserve our resources, reverse pollution, and so on. We will then survive as a civilization, though perhaps at a high cost and, in the end, rebuild what damage is done and progress onward for eons, perhaps, on a new and better basis.

In that case, we might argue that one out of every ten extraterrestrial civilizations would survive and persist for extended periods.

We might therefore maintain that out of the 50,000,000 extraterrestrial civilizations that come into being, 5,000,000 would be long-lived.

In other words, even a pessimistic assessment of our own future does not altogether eliminate the possibility that there are millions of extraterrestrial civilizations now in existence, each of them advanced far beyond us in technology.

This brings us back to our original question: Where is everybody?

3. *It could be that there's no practical way of visiting us.*

If there are, let us say, 5,000,000 long-lived extraterrestrial civilizations in the Galaxy right now, that means that about 1 out of every 30,000 stars in our galaxy shines down on a long-lived extraterrestrial civilization. In the regions of the Galaxy where such civilizations are likely to occur, the average separation between stars is 9.2 light-years. The average separation between extraterrestrial civilizations would then be 9.2 times the cube root of 30,000, or 285 light-years.

If we take 285 light-years as the most probable distance to even the nearest extraterrestrial civilization, you can see that visiting us would be quite a chore.

After all, as far as we know now, the speed of light is the absolute limit as far as the speed of transportation or communication is concerned. The round trip from one civilization to its neighbor and back, allowing for the need to accelerate and decelerate, would surely be, at least, 1000 years.

Even if it is not a matter of traveling, but communication by radio or its equivalent, the time lapse for any exchange of informations between two neighboring civilizations would take, on the average, 570 years. All this could well be considered more trouble than it was worth.

There have been speculations, to be sure, that the speed-of-light limit can be circumvented by the use of tachyons or of black holes, but such things are entirely speculative, and there is a strong temptation to suppose that while millions of extraterrestrial civilizations might exist, each might be confined, both by natural law and by choice, to their own planetary systems.

In fact, we may again use an argument in reverse: (A) There are undoubtedly millions of long-lived civilizations in the Galaxy; (B) none have ever come to visit us; (C) the speed of light is an absolute limit and will never be circumvented.

Yet I would be hard put to it to answer those who maintain that an advanced technological civilization is bound to find an answer to the speed-of-light limitation. Just because *we* don't see the answer means nothing. A primitive society speculating on the possibility of communicating with other societies thousands of miles away might carefully calculate the endurance required of runners and the time it would take to complete the round trip; or it might consider the loudness required of drum signals to allow them to be heard at so great a distance; and it might come to the conclusion that there was no practical way, even in theory, in which such communication could be established. It would never conceive of either jet planes or radio.

Well, then, assume that the advanced technologies can do easily what we cannot even imagine and that they find it neither too troublesome nor too expensive to travel among the stars.

Then where is everybody?

We might argue that as the intelligences come out of their planetary systems, they war on each other to the point of mutual destruction.

This does not seem likely. If they are that warlike, they would not have survived to reach the stage of interstellar travel. If they had managed to survive to that stage and were nevertheless warlike, the first and most advanced technology would most likely have destroyed all other burgeoning civilizations, colonized all suitable planets, and a one-species Galactic Empire would be now in existence.

The very fact that we are here, developing without outside interference, would indicate there is no such warlike and conquering faster-than-light civilization.

Perhaps, then, it is just that they haven't found Earth as yet. After all, the Universe is huge.

That isn't likely, either. The Galaxy is 15,000,000,000 years old and the oldest technological civilization could be 10,000,000,000 years old. In 10,000,000,000 years, such a civilization would have had ample time to study every Sunlike star in the Galaxy. The likelihood of their missing even one is not worth considering.

It could be, then, that the extraterrestrial intelligences are humane beings that do not war on each other and that value life. They may know of our existence and are deliberately refraining from interfering with us in order that we may develop freely and, in due time, join a Galactic Federation.

This, to me, is the most attractive possible answer to the problem of non-visits and, for that very reason, it may be only wishful thinking.

One last point, however. Perhaps this whole argument is pointless because representatives of extraterrestrial civilizations *have* visited us in the past, as Erich von Däniken claims, and are continuing to visit us in the present, as the flying saucer enthusiasts insist.

I don't say this is inconceivable, but the silly bits of "evidence" presented by the credulous visitors-from-outer-space freaks are, so far at least, totally unworthy of consideration.

My own feeling, as I look over the list of possibilities, is that the most likely answer is the unbeatable existence of a speed-of-light limit. I suspect that the Galaxy is full of advanced but forever isolated civilizations.

But can we at least establish some sort of contact with them?

If visiting is out of the question and if regular and intimate exchanges of information are not in the cards, surely occasional contacts may be possible. It might be just something to let them know we're here or just something to find out they're there. No more than that, perhaps. Just "Hello" in either direction.

Why?

We might want to bring ourselves to their attention just as a matter of pride. You, as an individual, might want the world to know that you exist, so we, as a species, might well want the Galaxy to know that we exist.

We can do this by sending out a material message. In fact, we have already done so. On March 3, 1972, Pioneer 10 was launched. Its primary purpose was to pass near Jupiter and study it—which it did, most successfully, in December 1973. It picked up energy as it rounded Jupiter and then passed on through the outer regions of the solar system. By 1984 it will coast past Pluto's orbit and into the vastness of interstellar space.

Attached to Pioneer 10 is a gold-anodized aluminum plate 15 by 22.5 centimeters (6 by 9 inches), which carries a representation of two human beings and information that, if deciphered, will indicate the planetary system from which it originated and the time at which it originated. In 1973 a duplicate plaque was sent out when Pioneer 11 was launched, and in 1977, recordings of a vast mix of Earth sounds were sent out on Voyager 1 and 2.

These are not particularly effective means of communication. It will take Pioneer 10, for instance, about 80,000 years to reach Alpha Centauri—if it were aimed in that direction, which it isn't. Pioneer 10 will not, in fact, come close to any star we can see, close enough to enter its planetary system, for at least 10,000,000,000 years. The chances are thus enormously against anyone ever coming across those messages during the lifetime of humanity.

But why send material objects, when we can send radiation? We can send a beam of laser light or of microwaves and aim them directly at stars in whose planetary systems we believe that advanced civilizations may exist. It would take such radiation only 4.4 years to reach Alpha Centauri, not 80,000 years, and it would also involve a good deal less trouble and expense.

However, why send out a beam blindly?

Why not first listen before trying to shout? Other civilizations more advanced than we could send out more effective signals than we could. Why not, then, try to detect the existence and exact site of a specific extraterrestrial intelligence and *then* beam a message to that site?

Will it do us any good to detect some extraterrestrial intelligence?

I think so. Let me give you one example of the good it might do. The mere fact that one exists, especially if what we detect can tell us that it is farther advanced technologically than we are, will prove to us that it is possible for a species to develop an advanced civilization and yet not commit suicide. I, for one, would be delighted to learn that.

But then, suppose we don't succeed? Does that mean that all advanced technologies do commit suicide?

Not necessarily. It might mean that we simply haven't developed detecting systems delicate enough, or that we aren't listening at the right "keyhole" or that we are receiving the evidence but aren't intelligent enough or advanced enough to grasp the fact that they are communicating something.

It could be, then, that even though we build enormously complex and enormously expensive devices to detect signals from extraterrestrial intelligences, we may not succeed. It is even, on the whole, much more likely that we won't succeed than that we will.

The businessman I mentioned in my introduction might shake his head at this. How can you begin such a task unless you are sure, in advance, that you will succeed?

The answer to that is that, in one way, we can't fail. Even if we detect no signals of extraterrestrial life, the complex detecting systems we build will surely extract so much other information from the sky generally as to vastly increase our understanding of the Universe we live in—and that can only be good.

16. The Road to Infinity

When I sold my first science fiction story just forty years ago,* my father, terribly impressed at finding himself in the same family with an important literary figure, said to me, "You must now behave with great dignity, Isaac."

I burst into laughter when he said it.

For all I knew at that time, my father was right and writers were supposed to develop the facial attitudes of stuffed frogs, but I knew I never would. I was very young at the time,† but I had already had occasion to examine my psyche well enough to know that I lacked even a chemical trace of dignity and was not likely ever to develop any.

And I haven't done so, either.

That makes it hard for people to treat me with the proper awe and reverence, so that when my good friend and official photographer of all things science-fictionish Jay Kay Klein wished to know how large a black hole would be if made by objects of ordinary mass, he put it this way:

"How large would Isaac Asimov be if he were compressed into a black hole?"‡

Another good friend of mine, the physicist and long-time s.f. fan Milton Rothman, offered to answer the ques-

---

* Not bad for someone who's only a little over thirty.
† A little over minus ten, obviously.
‡ For black holes, see "Final Collapse," in my Doubleday collection *Quasar, Quasar, Burning Bright.*

tion. To answer it, he had to know my mass, and, making an unfortunately correct guess, he handed Jay Kay the answer.

The thought that I might object at being compressed into a black hole deterred neither Jay Kay nor Milt. Neither one would have dared to do that to my fellow science-fiction writers L. Sprague de Camp or Robert A. Heinlein, both of whom have natural dignity, nor to Harlan Ellison, who would take them apart if they tried. But no one worries about good old Isaac.

However, I've lost twenty pounds since that classic exchange, so the figures have altered, and I'd like to start all over again—from the beginning, of course.

Any black hole has a "Schwarzschild radius," named for the German astronomer Karl Schwarzschild, who first determined its value long before black holes had become a routine topic for cocktail party conversations. (Schwarzschild died in 1916.) The Schwarzschild radius is the distance from the center of a black hole to a point on an imaginary sphere surrounding it where the escape velocity is equal to the speed of light.

Nothing can escape from the black hole (a statement I'll modify later in this chapter) once it has passed closer to the center than the Schwarzschild radius. That, according to the conventional point of view, is the point of no return. We will therefore call this the radius of the black hole, and twice the Schwarzschild radius is the diameter of the black hole.

The way you can calculate the diameter of a black hole is to make use of the equation, $D = 4GM/C^2$, where D is the diameter of the black hole, G is the gravitational constant, C is the speed of light, and M is the mass of the black hole.

The gravitational constant and the speed of light have well-known values that are fundamental and that scientists assume to be the same everywhere in space and everywhen in time. Making use of a system of units that will give us an answer in meters (and there's no need to specify the units if you'll just trust me), the value of G is 6.670×10^{-11}, while the value of C is 2.998×10^8.

The value of $4G/C^2$ is therefore equal to 2.97×10^{-27} and the equation for the diameter of a black hole becomes $D = 2.97 \times 10^{-27} M$.

As it happens, my own present mass is 74.8 kilograms. Substituting 74.8 for M in the equation, we find that if I were compressed into a black hole, I would have a diameter of 2.22×10^{-25} meters.

It is very difficult to visualize how small that is. It would take 10,000,000,000 objects the size of myself as a black hole to stretch across the diameter of a single proton.

Working with such small objects is clearly unrealistic. Let's consider astronomical bodies instead and ask what diameter each would have if its mass were compressed into a black hole.

Suppose we start with the Moon which is several billion trillion times as massive as I am; then the Earth, which is 81 times as massive as the Moon; then Jupiter, which is 318 times as massive as the Earth; then the Sun, which is 1048 times as massive as Jupiter; then the globular cluster in Hercules, which has a mass of about 1,000,000 times the Sun's; then our Galaxy, which is 150,000 times the mass of the Hercules globular cluster; finally our Universe, which is perhaps 100,000,000,000 times as massive as our Galaxy.

The results are in Table 9.

Exponential numbers are easy to manipulate but not necessarily easy to picture at a glance, so let's look at those black-hole diameters in another way.

To begin with, let's use a convenient terminology. Instead of saying "a black hole with the mass of the Moon," let's just say "B-Moon" and treat the other objects in the same way. The diameter of the B-Moon is about 1/5 millimeter, which would make it just about large enough to see without a magnifying glass, but it would be an object

TABLE 9

Object	Mass (kilograms)	Diameter of black hole (meters)
Moon	7.35×10^{22}	2.18×10^{-4}
Earth	5.98×10^{24}	1.78×10^{-2}
Jupiter	1.90×10^{27}	5.64
Sun	1.99×10^{30}	5.91×10^3
Globular cluster	2×10^{36}	6×10^9
Galaxy	3×10^{41}	9×10^{14}
Universe	3×10^{52}	9×10^{25}

with the full mass of the Moon, which is the amazing thing about it.

The B-Earth would be 1 3/4 centimeters across (about 7/10 inch)—the size of a marble.

With the B-Jupiter, we are beginning to get somewhere, for it would be 5.64 meters (18.5 feet) across. It would fill a good-sized, two-story-high living room.

The B-Sun would be 5.91 kilometers (3.67 miles) across and would have the volume of a small asteroid.

The B-Cluster would be 6,000,000 kilometers (3,700,000 miles) across and would have nearly 80 times the volume of the Sun.

The B-Galaxy would be roughly 1,000,000,000,000 kilometers across, or about a tenth of a light-year, and would be far wider than the orbit of Pluto.

The B-Universe would be 10,000,000,000 light-years across, a very respectable size indeed.

You'll notice, from the equation cited above, that the diameter of a black hole is proportional to its mass, or, which is the same thing, that the mass is proportional to the diameter.

This is an odd thing and does not fit what we would expect of ordinary objects.

We know from geometry that the volume of a sphere is proportional to the cube of its diameter. In other words, the volume of a sphere that is 2 meters across is $2\times2\times2$, or 8, times as great as the volume of a sphere that is 1 meter across. (This is also true of cubes or of any three-dimensional object of any shape, as long as it does not change its shape or its proportions as it grows larger and smaller.)

If we imagine a sphere to be made of a substance of a certain density and if this density does not change as the sphere is made larger or smaller, then the mass of the sphere is proportional to the volume. If a large sphere has 8 times the volume of a small sphere, it also has a mass 8 times the mass of a small sphere.

Consequently, provided density is held constant, the mass of a sphere is proportional to the cube of its diameter; or, to put it in reverse, the diameter of a sphere is proportional to the cube root of its mass.

How, then, in a black hole, can it be that the diameter is not proportional to the cube root of its mass—but is proportional to the mass directly?

The answer is that density is *not* held constant in the case of a black hole. A large black hole is less dense than a small one; less mass is squeezed into a large black hole than you would expect from its volume and for that reason a black hole 2 meters (6.6 feet) across is not 8 times as massive as a black hole 1 meter (3.3 feet) across, but is only 2 times as massive.

To have this make sense consider what happens as any object is compressed—

The gravitational pull of the Earth on you, when you stand on its surface, is, let us say, 70 kilograms (154 pounds). This pull increases as the distance between you and the center of the Earth decreases, *provided* all the mass of the Earth stays between you and the Earth's center.

If you were to try to approach the Earth's center by burrowing through the Earth's crust and mantle, you would be leaving more and more of the Earth's mass on the other side of yourself. The mass on the other side would counter the Earth's pull, and you would actually experience a smaller gravitational pull as you burrowed.

When you reached the Earth's center, you would experience no pull at all and would be at zero gravity.

If, however, the Earth were compressed into a smaller and smaller sphere, with none of its mass lost (meaning that its density were increasing steadily), and if you remained on its surface while this was happening, you would be approaching the center steadily, with all the mass of the Earth remaining between you and the center. The gravitational pull on you would therefore increase and carry you forward on the road to infinity, for once the Earth had been compressed to zero volume and infinite density, and you were at the infinitely-dense center, the gravitational pull on you would be infinite, too.

Somewhere in the course of that compression, the surface gravity would reach the stage where the escape velocity would equal the speed of light and that would mark the Schwarzschild radius.

This would be true for any body possessing mass, however small that mass might be.

Naturally, the more massive a body, the greater the surface gravity to begin with and the less it need be compressed to achieve a surface gravity large enough to produce an escape velocity equal to the speed of light. Since the more massive body need be compressed less, it reaches a lesser density level when it becomes a black hole. Suppose, for instance, we calculate the density of the various black holes we talked about earlier, as in Table 10.

TABLE 10

Black hole	Density (kilograms per cubic meter)
B-Moon	4.2×10^{34}
B-Earth	6.4×10^{30}
B-Jupiter	6.3×10^{25}
B-Sun	5.8×10^{21}
B-Cluster	5.6×10^{7}
B-Galaxy	2.5×10^{-8}
B-Universe	2.5×10^{-25}

We have grown accustomed to think of black holes as being extremely dense, and that is reasonable if we think of black holes that possess masses no greater than individual stars. The density of water is 1000 kilograms per cubic meter so that the B-Sun is a billion billion times as dense as water.

Black holes of substellar size are denser still. The B-Moon is 10,000,000,000,000 times denser than the B-Sun, while the density of the B-Asimov would be 1.6×10^{77} kilograms per cubic meter. At *that* density, the entire Universe could be fitted into the volume of an ordinary atom.

But what if we consider black holes of superstellar size? The B-Cluster is only 100,000 times as dense as water and the B-Galaxy is far *less* dense than water.

As a matter of fact, the B-Galaxy is about 1/500 as dense, on the average, as air is on the Earth's surface. Of course, within the tenth-of-a-light-year span of the B-Galaxy, the density may well not be uniform throughout. I could easily imagine that it grows steadily denser as the center is approached. If that is so, then the regions outside the central regions are all the less dense to make up for it and, in short, the major part of such a black hole must strike us, in terms of density, as a pretty good vacuum.

This is even more so in the case of the B-Universe, where the diameter is not very much smaller than the diameter of the actual Universe, and the average density is not very much greater than what it is in the actual Universe.

Well, then, what if the Universe has a somewhat larger mass than I have estimated. Astronomers, at the moment, think not; but astronomers may be wrong. There is some indication that galaxies might be surrounded by hazes of stars and gas outside their clearly observable limits and this may add unexpected mass to those galaxies and to the Universe as a whole. There may also be more and larger black holes in existence than astronomers count upon, and so on.

Suppose, then, that the mass of the Universe is a hundred times larger than I had estimated earlier in the article. In that case, the B-Universe would have a diameter of 9×10^{27} meters, or 1,000,000,000,000 light-years, some 40 times the diameter of the Observable Universe. The average density of the B-Universe would, in the enlarged case, be about 2.5×10^{-29} kilograms per cubic meter, which is just about what the average density of the actual Universe may be.

In other words, if the Universe were somewhat more massive than astronomers think it is, then it would *be* a black hole and that would account for a great many things.

For instance, if the Universe had more mass than astronomers now think, it would not expand forever. The over-all gravitational pull would gradually bring expansion to a stop and a very slow contraction would begin and then slowly accelerate.

If the Universe were a black hole, you could see why this would be. None of it could move outward beyond the Schwarzschild radius. That would represent the maximum diameter it could reach by expansion, and when that is reached (or perhaps even before it is), contraction must start.

During the period of expansion we are now undergoing, it is possible that local conditions at the centers of galaxies, at the centers of globular clusters, at the centers of massive stars may produce black holes of considerably less mass and size than that of the B-Universe. These smaller black holes (within the B-Universe) have masses that range from perhaps 3 times that of the Sun to perhaps 10,000,000,000 times that of the Sun. All are comparatively small, with diameters ranging from that of a small asteroid to that of a large planetary system. The matter within such black holes is packed in at, generally, enormous densities and this has limited freedom of expansion within the black hole.

During the period of contraction of the Universe, additional small black holes will form and a larger and larger percentage of the matter of the Universe will be pent up in close quarters, unable to break out past the restrictive bondage of their various Schwarzschild radii.

Eventually, though, the contraction will push the black

holes together into one large black hole with the mass of the Universe.—But that is not a stable situation. The one large B-Universe has its Schwarzschild radius 500,000,-000,000 light-years away in every direction. It has remained there all through the great cycle of expansion and contraction.

There would then be enormous room in which to expand and the "cosmic egg," as soon as it forms (heating up, as it coalesces, to an unimaginably high temperature), promptly bounces outward again in an incalculably vast explosion and the whole thing starts over again.

The British physicist Stephen Hawking has applied quantum mechanics to the relativistic equations used to work out the properties of black holes, and it turns out that the usual notion that nothing at all can ever leave a black hole must be slightly modified.

The energy of a rotating black hole is sometimes converted into a particle/antiparticle pair at the Schwarzschild radius. The two particles of the pair move off in opposite directions. One will move into the black hole, but the other will move away from it and escape. The result is a slow drizzle of mass *out* of the black hole, accompanied by electromagnetic radiation. The effect is that the black hole undergoes a kind of evaporation.

This happens at the surface of the black hole only. The more massive the black hole, the larger it is and the smaller the percentage of its total volume that is near the surface. A very small black hole has almost all its substance quite close to the surface so that almost all of it is subject to the evaporating effect. A very large black hole has almost all of its substance quite far inside the surface so that very little of it is subject to the evaporating effect.

One way of looking at this is to suppose that black holes have a temperature and are therefore boiling away, so to speak. The smaller they are, the hotter they are and the faster they boil away. In fact, the quantum effects cause black holes to radiate mass at a rate equivalent to what would happen if they were at a temperature of $10^{23}/M$ degrees above absolute zero, where M is the mass of the black hole in kilograms.

A B-Jupiter would, therefore, behave as though it were at a temperature of 0.0005° K. It would be only a two-thousandth of a degree above absolute zero and it would take an exceedingly long time to evaporate away. Anything more massive still would be still closer to zero and the evaporation would be at such a small rate that it could, in all conscience, be ignored.

This is all the more true since the larger a black hole, the more likely it is to encounter matter it can absorb. The more massive a black hole, then, the more likely it is to grow and become still more massive. The B-Jupiter, for instance, is bound to pick up matter at a rate great enough to replace what it loses by evaporation many times over, so that its temperature would drop steadily lower and there would be no question of its disappearance until such time (if ever) as it melts into the cosmic egg. Despite the quantum-effect correction, we can still look at such black holes as permanent, and we can still say that nothing emerges and be only negligibly wrong in saying so.

What about black holes smaller than the B-Jupiter? The B-Earth is at a temperature of 0.016° K., still within only a sixtieth of a degree of absolute zero, and the B-Moon has a temperature of 1.4° K. Even the B-Moon doesn't evaporate much, but if we get black holes that are smaller still—

But wait! Where are all these small black holes going to come from? The only process we know of that will form a black hole is the explosion of a giant star into a supernova and that will result in a black hole somewhat more massive than our Sun. At the center of giant galaxies, a black hole, originating from one star, may grow, through accretion of matter, the swallowing of stars whole, or the coalescence with other black holes, to masses of even some billions of times that of our Sun. Eventually, the whole Universe may melt, momentarily, into a cosmic egg. —But where in all this can we get *small* black holes?

As far as we know, considering the processes that go on in the Universe today, we can't even get something as small as a B-Sun, let alone B-objects smaller still.

Stephen Hawking, however, suggested an entirely different mechanism for black-hole formation, one that can't take place now.

At the time of the big bang, he supposes, the totally indescribable fury of the explosion would set up local pressures, here and there, that would be greater than any now existing at the center of the most massive and densest objects. Some bits of matter would be pressed together, in consequence, toward zero volume and infinite density and, on that road to infinity, would become black holes. Any quantities of matter, even quite insignificant quantities, might form black holes in this way.

Hawkins calls such small black holes "mini-black holes." They would have formed only at the time of the big bang and never since. Any that exist now are as old as the present Universe, and Hawking speculates that there might be as many as 300 of them per cubic light-year.

They would come in all sizes and some would be so small that their effective temperature would be quite high and their radiation rate quite appreciable. A black hole with the mass of the large asteroid Ceres would have an effective temperature of 8.5° K., and one with the mass of the small asteroid Icarus (which has a mass of merely 5,000,000,000,000 kilograms) would have an effective temperature of 20,000,000,000° K. By the time we get down to the small B-asteroids, they're evaporating at an appreciable rate. (A B-Asimov would have an effective temperature of a billion trillion degrees K.)

These smaller mini-black holes could be evaporating faster than they accrete matter, and in that case they would not last forever. If we neglect the accretion of matter, the lifetime of a black hole is $10^{-19}M^3$ seconds, where M is the mass in kilograms.

If a B-Sun accreted no matter, it would take something like 3×10^{64} years for it to evaporate its mass away—

which may not be eternity, but which I'm willing to treat as a practical equivalent, especially since it *will* accrete matter.

On the other hand, a B-Asimov would evaporate so rapidly that it would last only a little over a hundred-trillionth of a second. There would therefore be no use in compressing me into a black hole. I would explode instantly in a micro-small version of the big bang.

Well, then, how small must a mini-black hole be to have a lifetime equal to 15,000,000,000 years, which is the length of time (more or less) since the big bang? That length of time, the present age of the Universe, is 4.73×10^{17} seconds. We must then say that $10^{-19}M^3 = 4.73 \times 10^{17}$ and solve for M.

It comes out to 1.68×10^{12} kilograms, and such a black hole in its natural state would be the equivalent of a spherical asteroid about 1 kilometer (0.62 miles) in diameter.

In other words, any mini-black hole with a mass less than that of a kilometer-diameter asteroid would not be around now. It would (assuming that it hadn't accreted matter in its lifetime) have vanished in the time that has elapsed since the big bang. The smaller it was when it formed, the longer ago it would have vanished.

Any mini-black hole with a mass more than that of a kilometer-diameter asteroid would still be around now, even if it had accreted no matter. What's more, the lifetime increases rapidly with increasing mass. If the mass were only 10 per cent greater than the kilometer-diameter asteroid, the mini-black hole would hang around for another 20,000,000,000 years.

And if a mini-black hole has just about the mass of a kilometer-diameter asteroid and has added no mass to itself during its lifetime, it should be coming to the end of its life as a mini-black hole just about now.

What's more, it's not a quiet end. As a mini-black hole evaporates, its mass decreases. As its mass decreases, its effective temperature and its rate of evaporation increase.

In other words, the more a mini-black hole evaporates, the faster it continues to evaporate—and the faster—and the faster—until the last million kilograms go in ten seconds.

The final explosion (which, for the amount of matter that explodes, is very violent) results in the production of a shower of gamma rays with characteristics that Hawking's calculations pinpoint.

It is Hawking's suggestion that astronomers be on the lookout for gamma-ray showers of these specific characteristics. If such showers are detected, it would be very difficult to blame them on anything but the disappearance of a mini-black hole that was formed at the time of the big bang and that just happened to be massive enough to last till right now.

These gamma rays have not yet been detected, as far as I know, but they may be at any time in the future. And if they are, then we will know that at least some small bits of matter can (given enough time) retrace their passage on the road to infinity.

G LIFE AND DEATH

and my own guess is one Earthlike planet for every ten

17. The Subtlest Difference

Since I write on many subjects in these essays, and always with an insufferable air of knowledge and authority, it would probably do all my Gentle Readers a lot of good to have me own up to stupidity from time to time. I will gladly do so, since I have many examples to choose from.

About two weeks ago, for instance, I sat in the audience listening to a private detective talk about his profession. He was young, personable, intelligent, and a very good speaker. It was a pleasure to listen to him.

He told about the way in which he had helped get off an important wrong-doer by being able to show that the police had conducted an illegal search. He then explained that he felt perfectly justified in trying to get off people who were undoubtedly criminals because (1) they are constitutionally entitled to the best possible defense; (2) if the prosecution's tactics are faulty, the criminals will be released on appeal anyway; and (3) by insisting on due process to the last detail, we are protecting everybody, even ourselves, from a government that, without constant vigilance, can only too easily become a tyranny.

I sat there and nodded. Good stuff, I thought.

He then switched to humorous stories. One was about a professional man, separated from his wife and living with his secretary. Wanting to get rid of the secretary, the man asked the private detective to follow the secretary and let himself get caught at it by her. The secretary would then tell her lover she was being followed and he would say,

221

"Oh, my goodness, my wife is on my trail. We must split up."

Although the private detective did everything he could to be caught, the secretary refused to be disturbed about it and the little plan failed.

Now my hand went up and, out of sheer stupidity, I asked a silly question. I said, "I understand the constitutional issues involved in working on the side of criminals. What, however, is the constitutional issue involved in helping some guy pull a dirty, sleazy trick on some poor woman? Why did you do that?"

The private detective looked at me with astonishment and said, "He *paid* me."

Everyone else in the audience snickered and nudged each other, and I realized I was the only person there who was so stupid that he had to have that explained to him.

In fact, so clearly was I being snickered at, that I didn't have the courage to ask the next question, which would have been, had I dared ask it, "But if being a private detective lays you open to the temptation to do filthy jobs for the money it brings you, why don't you choose some other profession?"

I suppose there's a simple answer to that question, too, which I'm too stupid to see.

—And now, having warned you all of my inability to understand simple things, I will take up a very difficult matter indeed, the question of life and death. In view of my confession, nothing I say need be taken as authoritative or as anything, indeed, but my opinion. Therefore, if you disagree with me, please feel free to continue to do so.

What is life and what is death and how do we distinguish between the two?

If we're comparing a functioning human being with a rock, there is no problem.

A human being is composed of certain types of chemicals intimately associated with living things—proteins, nucleic acids, and so on—while a rock is not.

Then, too, a human being displays a series of chemical

changes that make up its "metabolism," changes in which food and oxygen are converted into energy, tissues, and wastes. As a result, the human being grows and can reproduce, turning simple substances into complex ones in apparent defiance of the second law of thermodynamics.* A rock does not do this.

Finally, a human being demonstrates "adaptive behavior," making an effort to preserve life, to avoid danger and seek safety, both by conscious will and by the unconscious mechanisms of his physiology and biochemistry. A rock does not do this.

But the human/rock contrast offers so simple a distinction between life and death that it is trivial and doesn't help us out. What we should do is take a more difficult case. Let us consider and contrast not a human being and a rock, but a live human being and a dead human being.

In fact, let's make it as difficult as possible and ask what the essential difference is between a human being just a short time after death—say, five minutes before and five minutes after.

What are the changes in those ten minutes?

The molecules are still all there—all the proteins, all the nucleic acids. Nevertheless, *something* has stopped being there, for where metabolism and adaptive behavior had been taking place (however feebly) before death, they are no longer taking place afterward.

Some spark of life has vanished. What is it?

One early speculation in this respect involved the blood. It is easy to suppose that there is some particular association between blood and life, one that is closer and more intimate than that between other tissues and life. After all, as you lose blood, you become weaker and weaker, and finally you die. Perhaps, then, it is blood which is the essence of life—and, in fact, life itself.

A remnant of this view will be found in the Bible, which in places explicitly equates life and blood.

---

* But only *apparent.* See "The Judo Argument," in my Doubleday collection, *The Planet That Wasn't.*

Thus, after the Flood, Noah and his family, the only human survivors of that great catastrophe, are instructed by God as to what they might eat and what they might not eat. As part of this exercise in dietetics, God says: "But flesh with the life thereof, which is the blood thereof, shall ye not eat" (Genesis 9:4).

In another passage on nutrition, Moses quotes God as being even more explicit and as saying, "Only be sure that thou eat not the blood: for the blood is the life; and thou mayest not eat the life with the flesh" (Deuteronomy 12:23). Similar statements are to be found in Leviticus 17:11 and 17:14.

Apparently, life is the gift of God and cannot be eaten, but once the blood is removed, what is left is essentially dead and has always been dead and may be eaten.

By this view, plants, which lack blood, are not truly alive. They do not live, but merely vegetate and serve merely as a food supply.

In Genesis 1:29–30, for instance, God is quoted as saying to the human beings he has just created: "Behold, I have given you every herb bearing seed, which upon the face of all the earth, and every tree, in the which is the fruit of a tree yielding seed; to you it shall be for meat. And to every beast of the earth, and to every fowl of the air, and to everything that creepeth upon the earth, wherein there is life, I have given every green herb for meat . . ."

Plants are described as "bearing seed" and "yielding seed," but in animals "there is life."

Today we would not make the distinction, of course. Plants are as alive as animals, and plant sap performs the functions of animal blood. Even on a purely animal basis, however, the blood theory would not stand up. Although loss of blood in sufficient quantities inevitably leads to loss of life, the reverse is not true. It is quite possible to die without the loss of a single drop of blood; indeed, that often happens.

Since death can take place when, to all appearances, nothing material is lost, the spark of life must be found in something more subtle than blood.

What about the breath then? All human beings, all animals, breathe.

If we think of the breath, we see that it is much more appropriate as the essence of life than blood is. We constantly release the breath, then take it in again. The inability to take it in again invariably leads to death. If a person is prevented from taking in the breath by physical pressure on his windpipe, by a bone lodged in his throat, by being immersed in water—that person dies. The loss of breath is as surely fatal as the loss of blood; and the loss of breath is the more quickly fatal, too.

Furthermore, where the converse is not true for blood, (where people can die without loss of blood) the converse *is* true for air. People cannot die without loss of air. A living human being breathes, however feebly, no matter how close he is to death; but after death, he does not breathe, and that is always true.

Furthermore, the breath itself is something that is very subtle. It is invisible, impalpable, and, to early people, it seemed immaterial. It was just the sort of substance that would, and should, represent the essence of life and, therefore, the subtle difference between life and death.

Thus, in Genesis 2:7, the creation of Adam is described thus: "And the Lord God formed man of the dust of the ground, and breathed into his nostrils the breath of life; and man became a living soul."

The word for "breath" would be *ruakh* in Hebrew and that is usually translated as "spirit."

It seems a great stretch from "breath" to "spirit," but that is not so at all. The two words are literally the same. The Latin *spirare* means "to breathe" and *spiritus* is "a breath." The Greek word *pneuma* which means "breath," is also used to refer to "spirit." And the word "ghost" is derived from an Old English word meaning "breath." The word "soul" is of uncertain origin, but I am quite confident that if we knew its origin, it, too, would come down to breath.

Because, in English, we have a tendency to use words of Latin and Greek derivation and then forget the meaning of the classic terms, we attach grandiosity to concepts that don't belong there.

We talk of the "spirits of the dead." The meaning would be precisely the same, and less impressive, if we spoke of the "breath of the dead." The terms "Holy Ghost" and "Holy Spirit" are perfectly synonymous and mean, essentially, "God's breath."

It might well be argued that the literal meaning of words means nothing, that the most important and esoteric concepts must be expressed in lowly words, and that these words gather their meaning from the concept and not vice versa.

Well, perhaps if one believes that knowledge comes full-blown by supernatural revelation, one can accept that. I think, however, that knowledge comes from below, from observation, from simple and unsophisticated thinking that establishes a primitive concept that gradually grows complex and abstract as more and more knowledge is gathered. Etymology, therefore, is a clue to the *original* thought, overlain now by thousands of years of abstruse philosophy. I think that people noticed the connection of breath and life in a quite plain and direct way and that all the subtle philosophical and theological concepts of spirit and soul came afterward.

Is the human spirit as formless and impersonal as the breath that gave it its name? Do the spirits of all the human beings who have ever died commingle into one mixed and homogenized mass of generalized life?

It is difficult to believe this. After all, each human being is distinct and different in various subtle and not-so-subtle ways from every other. It would seem natural, then, to suppose that the essence of each one's life has, in some ways, to be different from every other. Each spirit, then, would retain that difference and would remain somehow reminiscent of the body it once inhabited and to which it lent the property and individuality of life.

And if each spirit retains the impress that gave the body

its characteristic properties, it is tempting to suppose that the spirit possesses, in some subtle, airy, and ethereal manner, the form and shape of the human body it inhabited. This view may have been encouraged by the fact that it is common to dream of dead people as being still alive. Dreams were often given much significance in earlier times (and in modern times, too, for that matter) as messages from another world, and that would make it seem like strong evidence that the spirits resembled the bodies they had left.

For modesty's sake, if for no other reason, such spirits are usually pictured as clad in formless white garments, made of luminous cloud or glowing light, perhaps, and that, of course, gives rise to the comic-strip pictures of ghosts and spirits wearing sheets.

It is further natural to suppose a spirit to be immortal. How can the very essence of life die? A material object can be alive or dead, according to whether it contains the essence of life or not; but the essence of life can only be alive.

This is analogous to the statement that a sponge can be wet or dry depending on whether it contains water or not, but the water itself can only be wet; or that a room can be light or dark depending on whether the Sun's rays penetrate it or not, but the Sun's rays can only be light.†

If you have a variety of spirits or souls which are eternally alive and which enter a lump of matter at birth and give it life and then leave it and allow it to die, there must be a vast number of spirits, one for each human being who has ever lived or ever will live.

This number may be increased further if there are also

---

† You can argue both points and say that water at a temperature low enough to keep it non-melting ice, or water in the form of vapor, is not wet; and that the Sun's rays, if ultraviolet or infrared, are not light in appearance. However, I am trying to argue like a philosopher and not like a scientist—at least in this paragraph.

spirits for various other forms of life. It may be decreased if the spirits can be recycled—that is, if a spirit, on leaving one dying body, can then move into a body being born.

Both these latter views have their adherents, sometimes in combination, so that there are some people who believe in transmigration of souls throughout the animal kingdom. A man who has particularly misbehaved might be born again as a cockroach, whereas, conversely, a cockroach can be reborn as a man if it has been a very good and noble cockroach.

However the matter is interpreted, whether spirits are confined to human beings or spread throughout the animal kingdom or whether there is transmigration of souls or not, there must be a large number of spirits available for the purpose of inducing life and taking it away. Where do they all stay?

In other words, once the existence of the spirit is accepted, a whole spirit world must be assumed. This spirit world may be down under the earth or up somewhere at great heights, on another world, or on another "plane."

The simplest assumption is that the spirits of the dead are just piled up underground, perhaps because the practice of burying the dead is a very ancient one.

The simplest underground dwelling place of the spirits would be one that is viewed as a gray place of forgetfulness, like the Greek Hades or the Hebrew Sheol. There the situation is almost like a perpetual hibernation. Sheol is described as follows in the Bible: "There the wicked cease from troubling; and there the weary be at rest. There the prisoners rest together; they hear not the voice of the oppressor. The small and great are there; and the servant is free from his master" (Job 3:17–19). And Swinburne describes Hades in *The Garden of Proserpine,* which begins:

Here, where the world is quiet,
   Here, where all trouble seems
Dead winds' and spent waves' riot
   In doubtful dreams of dreams.

This nothingness seems insufficient to many people and a rankling feeling of injustice in life tempts them to imagine a place of after-death torture where the people they dislike get theirs—the Greek Tartarus or the Christian Hell.

The principle of symmetry demands the existence of abodes of bliss as well for the people they like—Heaven, the Islands of the Blest, Avalon, the Happy Hunting Grounds, Valhalla.

All of this massive structure of eschatology is built up out of the fact that living people breathe and dead people don't and that living people desperately *want* to believe that they will not truly die.

Nowadays, we know, of course, that the breath has no more to do with the essence of life than blood does; that it, like blood, is merely the servant of life. Nor is breath insubstantial, immaterial, and mysterious. It is as material as the rest of the body and is composed of atoms no more mysterious than any other atoms.

Yet despite this, people still believe in life after death—even people who understand about gases and atoms and the role of oxygen. Why?

The most important reason is that regardless of evidence or the lack of it, people still want to believe. And because they do, there is a strong urge to believe even irrationally.

The Bible speaks of spirits and souls and life after death. In one passage, King Saul even has a witch bring up the spirit of the dead Samuel from Sheol (1 Samuel 28:7–20). This is enough for millions of people, but many of our secular and skeptical generation are not really inclined to accept, undiscriminatingly, the statements present in the collection of the ancient legends and poetry of the Jews.

There is, of course, eyewitness evidence. How many people, I wonder, have reported having seen ghosts and spirits? Millions, perhaps. No one can doubt that they have made the reports; but anyone can doubt that they have actually seen what they have reported they have seen. I can't imagine a rational person accepting these stories.

There is the cult of "spiritualism" which proclaims the ability of "mediums" to make contact with the spirit world. This has flourished and has attracted not only the uneducated, ignorant, and the unsophisticated, but, despite the uncovering of countless gross frauds, even such highly intelligent and thoughtful people as A. Conan Doyle and Sir Oliver Lodge. The vast majority of rational people, however, place no credence in spiritualism at all.

Then, too, over twenty years ago there was a book published called *The Search for Bridey Murphy*, in which a woman was supposedly possessed by the spirit of a long-dead Irishwoman, with whom one could communicate if her hostess were hypnotized. For a while, this was advanced as evidence of life after death, but it is no longer taken seriously.

But then is there *any* evidence of life after death that can be considered as scientific and rational?

Right now, there are claims that scientific evidence exists.

A physician named Elizabeth Kübler-Ross has been presenting statements she says she has received from people on their deathbeds that seem to indicate the existence of life after death—and a whole rash of books on this subject are being published, each book, of course, being guaranteed large sales among the gullible.

According to these reports now coming out, a number of people who have seemed to be "clinically dead" for a period of time have nevertheless managed to hang on to life, to have recovered, and then to have told of their experiences while they were "dead."

Apparently, they remained conscious, felt at peace and happy, watched their body from above, went through dark tunnels, saw the spirits of dead relatives and friends, and in some cases encountered a warm, friendly spirit, glowing with light, who was to conduct them somewhere.

How much credence can be attached to such statements?

In my opinion, none at all!

Nor is it necessary to suppose the "dead" people are

lying about their experiences. A person who is near enough to death to be considered "clinically dead" has a mind that is no longer functioning normally. The mind would then be hallucinating in much the same way it would be if it were not functioning normally for any other reason—alcohol, LSD, lack of sleep, and so on. The dying person would then experience what he or she would expect to experience or want to experience. (None of the reports include Hell or devils, by the way.)

The life-after-deathers counter this by saying that people from all stations of life, and even from non-Christian India, tell similar stories, which lead them to believe there is objective truth to it.—I won't accept that for two reasons.

1. Tales of afterlife are widespread all over the world. Almost all religions have an afterlife, and Christian missionaries and Western communications technology have spread *our* notions on the subject everywhere.

2. Then, too, having experienced hallucinations of whatever sort, the recovered person, still weak perhaps and confused, must describe them—and how easy it must be to describe them in such a way as to please the questioner, who is usually a life-after-death enthusiast and is anxious to elicit the proper information.

All the experience of innumerable cases of courtroom trials makes it quite plain that a human being, even under oath and under threat of punishment, will, with all possible sincerity, misremember, contradict himself, and testify to nonsense. We also know that a clever lawyer can, by proper questioning, elicit almost any testimony from even an honest, truthful, and intelligent witness. That is why the rules of evidence and of cross-examination have to be so strict.

Naturally, then, it would take a great deal to make me attach any importance to the statements of a very sick person elicited by an eager questioner who is a true believer.

But in that case what about my own earlier statement that some change must have taken place in the passage from human life to human death, producing a difference that is not a matter of atoms and molecules?

The difference doesn't involve blood or breath, but it has to involve *something!*

And it does. Something was there in life and is no longer there in death, and that something *is* immaterial and makes for a subtle difference—the subtlest difference of them all.

Living tissue consists not merely of complex molecules, but of those complex molecules *in complex arrangement.* If that arrangement begins to be upset, the body sickens; if that arrangement is sufficiently upset, the body dies. Life is then lost even though all the molecules are still there and still intact.

Let me present an analogy. Suppose one builds an intricate structure out of many thousands of small bricks. The structure is built in the form of a medieval castle, with towers and crenellations and portcullises and inner keeps and all the rest. Anyone looking at the finished product might be too far away to see the small individual bricks, but he will see the castle.

Now imagine some giant hand coming down and tumbling all the bricks out of which the castle is built, reducing everything to a formless heap. All the bricks are still there, with not one missing. All the bricks, without exception, are still intact and undamaged.

But where is the castle?

The castle existed only in the arrangement of the bricks, and when the arrangement is destroyed, the castle is gone. Nor is the castle anywhere else. It has no existence of its own. The castle was created out of nothing as the bricks were arranged and it vanished into nothing when the bricks were disarranged.

The molecules of my body, after my conception, added other molecules and arranged the whole into more and more complex form, and in a unique fashion, not quite like the arrangement in any other living thing that ever lived. In the process, I developed, little by little, into a conscious something I call "I" that exists only as the arrangement. When the arrangement is lost forever, as it will be when I die, the "I" will be lost forever, too.

And that suits me fine. No concept I have ever heard, of either a Hell or of a Heaven, has seemed to me to be suitable for a civilized rational mind to inhabit, and I would rather have the nothingness.

THE BEST IN SCIENCE FICTION AND FANTASY FROM AVON ◭ BOOKS

URSULA K. LE GUIN

The Lathe of Heaven	43547	1.95
The Dispossessed	51284	2.50

ISAAC ASIMOV

Foundation	50963	2.25
Foundation and Empire	52357	2.25
Second Foundation	52290	2.25
The Foundation Trilogy (Large Format)	50856	6.95

ROGER ZELAZNY

Doorways in the Sand	49510	1.75
Creatures of Light and Darkness	35956	1.50
Lord of Light	44834	2.25
The Doors of His Face The Lamps of His Mouth	38182	1.50
The Guns of Avalon	31112	1.50
Nine Princes in Amber	51755	1.95
Sign of the Unicorn	53132	1.95
The Hand of Oberon	51318	1.75
The Courts of Chaos	47175	1.75

Include 50¢ per copy for postage and handling, allow 4-6 weeks for delivery.

Avon Books, Mail Order Dept.
224 W. 57th St., N.Y., N.Y. 10019

SF 2-81